Smart farms: Improving data-driven decision making in agriculture

It is widely recognised that agriculture is a significant contributor to global warming and climate change. Agriculture needs to reduce its environmental impact and adapt to current climate change whilst still feeding a growing population, i.e. become more 'climate-smart'. Burleigh Dodds Science Publishing is playing its part in achieving this by bringing together key research on making the production of the world's most important crops and livestock products more sustainable.

Based on extensive research, our publications specifically target the challenge of climate-smart agriculture. In this way we are using 'smart publishing' to help achieve climate-smart agriculture.

Burleigh Dodds Science Publishing is an independent and innovative publisher delivering high quality customer-focused agricultural science content in both print and online formats for the academic and research communities. Our aim is to build a foundation of knowledge on which researchers can build to meet the challenge of climate-smart agriculture.

For more information about Burleigh Dodds Science Publishing simply call us on +44 (0) 1223 839365, email info@bdspublishing.com or alternatively please visit our website at www.bdspublishing.com.

Related titles:

Advances in agri-food robotics
Print (ISBN 978-1-80146-277-8); Online (ISBN 978-1-80146-278-5, 978-1-80146-279-2)

Advances in crop modelling for a sustainable agriculture
Print (ISBN 978-1-78676-240-5); Online (ISBN 978-1-78676-242-9, 978-1-78676-243-6)

Advances in sensor technology for sustainable crop production
Print (ISBN 978-1-78676-977-0); Online (ISBN 978-1-78676-979-4, 978-1-78676-980-0)

Improving data management and decision support systems in agriculture
Print (ISBN 978-1-78676-340-2); Online (ISBN 978-1-78676-342-6, 978-1-78676-343-3)

Improving soil health
Print (ISBN 978-1-78676-204-7); Online (ISBN 978-1-78676-206-1, 978-1-78676-207-8)

Robotics and automation for improving agriculture
Print (ISBN 978-1-78676-272-6); Online (ISBN 978-1-78676-274-0, 978-1-78676-275-7)

Chapters are available individually from our online bookshop:
https://shop.bdspublishing.com

climate•SMART•publishing
IN AGRICULTURAL SCIENCE

BURLEIGH DODDS SERIES IN AGRICULTURAL SCIENCE
NUMBER 147

Smart farms: Improving data-driven decision making in agriculture

Edited by: Professor Claus Grøn Sørensen, Aarhus University, Denmark

burleigh dodds
SCIENCE PUBLISHING

Published by Burleigh Dodds Science Publishing Limited
82 High Street, Sawston, Cambridge CB22 3HJ, UK
www.bdspublishing.com

Burleigh Dodds Science Publishing, 1518 Walnut Street, Suite 900, Philadelphia, PA 19102-3406, USA

First published 2024 by Burleigh Dodds Science Publishing Limited

Library of Congress Control Number: 2024930829

British Library Cataloguing in Publication Data
A catalogue record for this book is available from the British Library

ISBN 978-1-80146-382-9 (print)
ISBN 978-1-80146-384-3 (PDF)
ISBN 978-1-80146-383-6 (ePub)
ISSN 2059-6936 (print)
ISSN 2059-6944 (online)

DOI: 10.19103/AS.2023.0132

Typeset by Deanta Global Publishing Services, Dublin, Ireland

Contents

Series list

Title	Series number
Achieving sustainable cultivation of maize - Vol 1 From improved varieties to local applications Edited by: Dr Dave Watson, CGIAR Maize Research Program Manager, CIMMYT, Mexico	001
Achieving sustainable cultivation of maize - Vol 2 Cultivation techniques, pest and disease control Edited by: Dr Dave Watson, CGIAR Maize Research Program Manager, CIMMYT, Mexico	002
Achieving sustainable cultivation of rice - Vol 1 Breeding for higher yield and quality Edited by: Prof. Takuji Sasaki, Tokyo University of Agriculture, Japan	003
Achieving sustainable cultivation of rice - Vol 2 Cultivation, pest and disease management Edited by: Prof. Takuji Sasaki, Tokyo University of Agriculture, Japan	004
Achieving sustainable cultivation of wheat - Vol 1 Breeding, quality traits, pests and diseases Edited by: Prof. Peter Langridge, The University of Adelaide, Australia	005
Achieving sustainable cultivation of wheat - Vol 2 Cultivation techniques Edited by: Prof. Peter Langridge, The University of Adelaide, Australia	006
Achieving sustainable cultivation of tomatoes Edited by: Dr Autar Mattoo, USDA-ARS, USA and Prof. Avtar Handa, Purdue University, USA	007
Achieving sustainable production of milk - Vol 1 Milk composition, genetics and breeding Edited by: Dr Nico van Belzen, International Dairy Federation (IDF), Belgium	008
Achieving sustainable production of milk - Vol 2 Safety, quality and sustainability Edited by: Dr Nico van Belzen, International Dairy Federation (IDF), Belgium	009
Achieving sustainable production of milk - Vol 3 Dairy herd management and welfare Edited by: Prof. John Webster, University of Bristol, UK	010
Ensuring safety and quality in the production of beef - Vol 1 Safety Edited by: Prof. Gary Acuff, Texas A&M University, USA and Prof. James Dickson, Iowa State University, USA	011
Ensuring safety and quality in the production of beef - Vol 2 Quality Edited by: Prof. Michael Dikeman, Kansas State University, USA	012
Achieving sustainable production of poultry meat - Vol 1 Safety, quality and sustainability Edited by: Prof. Steven C. Ricke, University of Arkansas, USA	013
Achieving sustainable production of poultry meat - Vol 2 Breeding and nutrition Edited by: Prof. Todd Applegate, University of Georgia, USA	014

Achieving sustainable cultivation of sorghum - Vol 1 031
Genetics, breeding and production techniques
Edited by: Prof. William Rooney, Texas A&M University, USA

Achieving sustainable cultivation of sorghum - Vol 2 032
Sorghum utilization around the world
Edited by: Prof. William Rooney, Texas A&M University, USA

Achieving sustainable cultivation of potatoes - Vol 2 033
Production, storage and crop protection
Edited by: Dr Stuart Wale, Potato Dynamics Ltd, UK

Achieving sustainable cultivation of mangoes 034
Edited by: Prof. Víctor Galán Saúco, Instituto Canario de Investigaciones Agrarias (ICIA), Spain and Dr Ping Lu, Charles Darwin University, Australia

Achieving sustainable cultivation of grain legumes - Vol 1 035
Advances in breeding and cultivation techniques
Edited by: Dr Shoba Sivasankar et al., formerly International Crops Research Institute for the Semi-Arid Tropics (ICRISAT), India

Achieving sustainable cultivation of grain legumes - Vol 2 036
Improving cultivation of particular grain legumes
Edited by: Dr Shoba Sivasankar et al., formerly International Crops Research Institute for the Semi-Arid Tropics (ICRISAT), India

Achieving sustainable cultivation of sugarcane - Vol 1 037
Cultivation techniques, quality and sustainability
Edited by: Prof. Philippe Rott, University of Florida, USA

Achieving sustainable cultivation of sugarcane - Vol 2 038
Breeding, pests and diseases
Edited by: Prof. Philippe Rott, University of Florida, USA

Achieving sustainable cultivation of coffee 039
Edited by: Dr Philippe Lashermes, Institut de Recherche pour le Développement (IRD), France

Achieving sustainable cultivation of bananas - Vol 1 040
Cultivation techniques
Edited by: Prof. Gert H. J. Kema, Wageningen University and Research, The Netherlands and Prof. André Drenth, University of Queensland, Australia

Global Tea Science 041
Current status and future needs
Edited by: Dr V. S. Sharma, formerly UPASI Tea Research Institute, India and Dr M. T. Kumudini Gunasekare, Coordinating Secretariat for Science Technology and Innovation (COSTI), Sri Lanka

Integrated weed management 042
Edited by: Prof. Emeritus Robert L. Zimdahl, Colorado State University, USA

Achieving sustainable cultivation of cocoa 043
Edited by: Prof. Pathmanathan Umaharan, Cocoa Research Centre – The University of the West Indies, Trinidad and Tobago

Robotics and automation for improving agriculture 044
Edited by: Prof. John Billingsley, University of Southern Queensland, Australia

Water management for sustainable agriculture 045
Edited by: Prof. Theib Oweis, ICARDA, Jordan

Improving organic animal farming 046
Edited by: Dr Mette Vaarst, Aarhus University, Denmark and Dr Stephen Roderick, Duchy College, UK

Introduction

The agricultural sector remains under increasing pressure to reduce its environmental impact and consequent contribution to climate change, while also producing enough food to feed a rapidly growing population. With the variety and volume of data, coupled with the advanced methods for data processing, a new era of digital agriculture is emerging as a possible solution to this monumental challenge.

This volume provides a comprehensive review of the recent advances in gathering and analysing data as a means of improving farm sustainability, productivity and profitability. The book discusses the evolution of farm information management systems, highlighting current trends and challenges, as well as methods of data acquisition and analysis, including the use of artificial intelligence.

Part 1 General

The first chapter of the book details trends in farm information management systems. Farm management information systems (FMIS) can assist farmers in their daily decision making and reporting, as well as in their communication with stakeholders. Chapter 1 begins by defining FMIS and their functional requirements and standard modular components. It then reviews the evolution of FMIS and the role of technical architecture models which help to solve interoperability and customisation challenges. The chapter concludes by reviewing future developments in digitalisation and how these are affecting the evolution of FMISs.

Chapter 2 explores the role of digital technologies in achieving sustainable agriculture. It begins by tracking the evolution of agriculture from its inception to the digital age, before moving on to discuss efficiency indicators in agriculture and the need to overcome inefficiency in agricultural production. Discussion then moves to the contrast between precision agriculture and digital agriculture, and the ways in which they can tackle inefficiency. A section analysing material and energy flows in agriculture is also included. Finally, digital agriculture and its role in the sustainability agenda is considered.

The next chapter considers key issues in incorporating proximal and remote sensor data into farm decision making. Chapter 3 presents five study cases on the application of smart technologies in farming and forestry systems. The first presents the results of a long-term study to calibrate a Grassmaster II capacitance probe to estimate pasture productivity in the Mediterranean Montado ecosystem. The second presents a methodology for identifying yield zones within an olive grove based on Sentinel-2 satellite data. The third case study provides an overview of different remote sensing sensors, data and methodologies used for estimating forest biomass. The fourth assesses the carbon footprint of horticultural crops such as potato, onion, carrot, melon and watermelon. The final case study focuses on precision livestock farming, specifically a mobile application focused on thermal comfort.

Chapter 4 concerns agri-semantics, developments to improve data interoperability to support farm information management and decision support systems in agriculture. An ontology-driven, knowledge-based framework for farm animal health management systems is proposed. The framework focuses on a cattle application scenario and provides a standardised framework, a species-specific livestock health ontology, resource descriptive framework data generation, and semantic interoperability. The chapter analyses the background knowledge, presents the methodology as a case study, and concludes with future research directions and challenges.

Chapter 5 discusses using support vector machines in agriculture. The chapter begins by defining the support vector machine model, covering hard- and soft-margin, and primal and dual support vector machines, as well as the use of the kernel trick in data classification. Next, developments in support vector machine algorithms are examined, including: sub-gradient descent, regularisation and stability, hinge loss, target functions, and multiclass and Bayesian support vector machines. Following this, the challenges in using support vector machines are considered, covering: long training time, sensitivity to missing values, difficulties in handling large-scale multi-class problems, and the complexities of adjusting parameters. Solutions to these shortcomings are then described in the form of other models, including: random forest, gradient boosting trees, and deep learning models. Finally, four case studies are highlighted: identifying damage from cotton spider mites, identifying trajectories of a wheat harvester, tractor emission prediction, and spring wheat yield prediction.

Part 2 Case studies

The first chapter of Part 2, Chapter 6 covers developing decision support systems for irrigation on farms. It begins by reviewing software engineering guidelines that lead to developing and easily maintained software tools. Water in the farm and the basic principles involving transfer of water in the farm are then explored with a special focus on irrigation scheduling. This is followed by examples of successful decision aids from the first generation all the way through the fourth. Finally, potential outcomes of fourth industrial revolution technologies are discussed, including the beginning of a complex system approach to decision making aids.

Chapter 7 highlights advances in crop disease forecasting models. The chapter starts by examining how to model complex crop-disease-environment dynamics. It then explores advances in crop disease forecasting from the use of precision agriculture surveillance and detection technology, and the use of Earth observation satellites to automate the collection of digital agriculture 'big' data. This is followed by a section focused on novel artificial intelligence-based methodologies. Finally, a case study is described, covering how greater capability for conducting science in the cloud, multidisciplinary collaboration, and early-warning systems are creating new opportunities and challenges for identifying, mitigating, and responding to disease outbreaks.

Finally, Chapter 8 focuses on the Australian experience of smart farming in extensive livestock production. The chapter begins by tracking the evolution from precision agriculture to smart farming, as well as the adoption of digital agriculture and smart farming. Next, the role of the Internet of Things in extensive livestock production systems, and technologies for livestock identification and tracking are explored. The chapter then moves on to describe how the Internet of Things can be applied in animal health and welfare, as well as covering combining Internet of Things devices, such as: liveweight and yield mapping, water intake and water infrastructure failure alerting, and feed base and stocking rate management. Finally, the rise of demonstration smart farms for extensive livestock production is reviewed.

Part 1

General

Chapter 1

Trends in farm information management systems

Liisa Pesonen, Natural Resources Institute (LUKE), Finland

1 Introduction

The work of a farm manager includes numerous decisions to make every day, with the number of decisions now growing with the deployment of information-intensive precision agriculture. Farm managers need to make decisions at strategic, tactical and operational levels. Decisions are needed also while executing tasks and when evaluating the success of completed tasks. The more information a farm manager has, the better the decisions usually are. The farming environment is very complex, combining biological, technological, economic, ecological and social processes. Each farm is individual in terms of resources, business models and objectives. Figure 1 illustrates a farm manager's information needs and challenges when running the farm. The rich picture provides a detailed representation of the complexity encompassed in farm management. Farms are encountering increasing pressures in information exchange with stakeholders, i.e. meeting reporting requirements for government environmental and economic support schemes, as well as responding to quality requirements from other food system actors (e.g. food processors and retailers) and consumers. The range of information requirements is expanding from information to meet safety and quality standards to information relating to performance in areas such as environmental impact and socially responsible working practices.

http://dx.doi.org/10.19103/AS.2023.0132.01

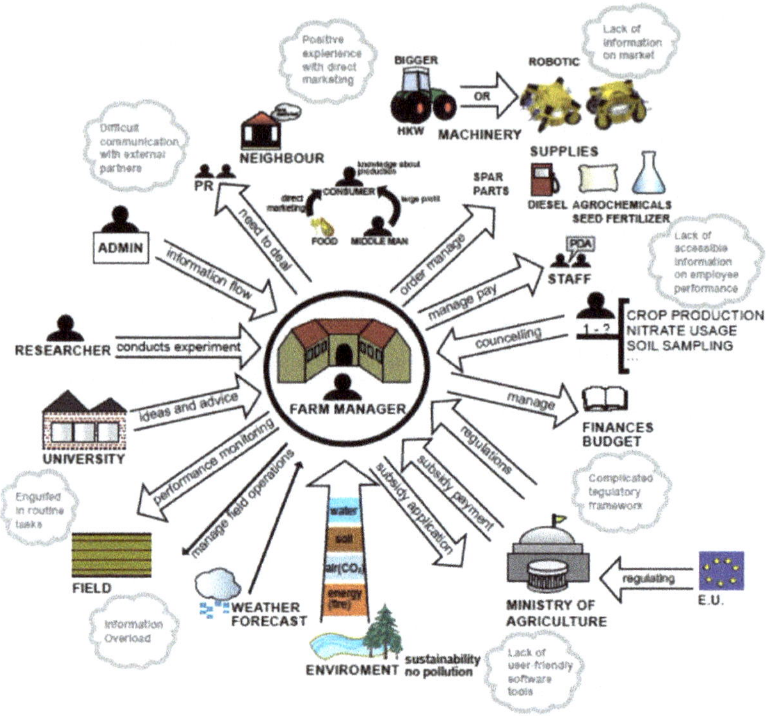

Figure 1 Information exchange needs (arrows) and challenges (clouds) of a farm manager presented in Sørensen et al. (2010).

Societies and food systems are digitalizing fast, and digital tools are also reaching farms in the form of precision agriculture methods and smart farming technologies. Some farms find the growing range of digital tools and the growing volume and range of potential information overwhelming. On the other hand, digital tools are supposed to assist in farming tasks through automated (and even autonomous) assistance, by measuring and recording production process events and environmental status both automatically and (potentially) continuously as electronic data. Data can then be used for more accurate decision-making and reporting. This creates the need for digital farm management information systems (FMIS). As Fig. 1 shows, the group of stakeholders who can potentially make use of FMIS is broad, including farm managers and workers, machine manufacturers and other technology providers, government agencies, banks and other financial institutions, food processors, farm input suppliers, advisory services and sometimes the research community. In an ideal situation, FMIS can assist farm managers in information exchange with all stakeholders they deal with. This chapter focuses on FMIS for open-field farming.

2 Defining farm management information systems

Sørensen et al. (2010) defined FMIS as 'a planned system for the collecting, processing, storing and disseminating of data in the form of information needed to carry out the operations functions of the farm'. The definition refers to terminology used in industry, where the role of a management information systems (MIS) is to analyse other systems that handle operational level activities in the organisation. The operations may include automation and human decision-making that require supporting information inputs and recording. An MIS is part of an overall information system (IS) that involves people, documents, technologies and procedures. It assists in business processes and tasks for producing services and products. MIS and IS are parts of enterprise resource planning (ERP) which is an enterprise-wide and cross-functional system coordinating all the resources, information and activities needed to complete business processes in a farm system (Fig. 2).

Globally, the majority of farms are small or medium enterprises (SMEs) (Van der Ploeg, 2016). They are often family farms where the farm manager carries out farm tasks (from field operations to business processes) with the help of family members or only a few workers. The farm manager needs decision support in several tasks regarding both field operations and business during the working day.

The latest developments in smart farming technologies, including cloud computing, mobile connections, and artificial intelligence (AI)-aided functionality (e.g. in data analysis), are making it possible to extend FMIS to include additional functionality so that they more closely resemble ERP systems (Henriyadi et al., 2022; Paraforos et al., 2016; Burlacu et al., 2014). Zysk et al. (2022) suggest renaming a FMIS as an agriculture information management

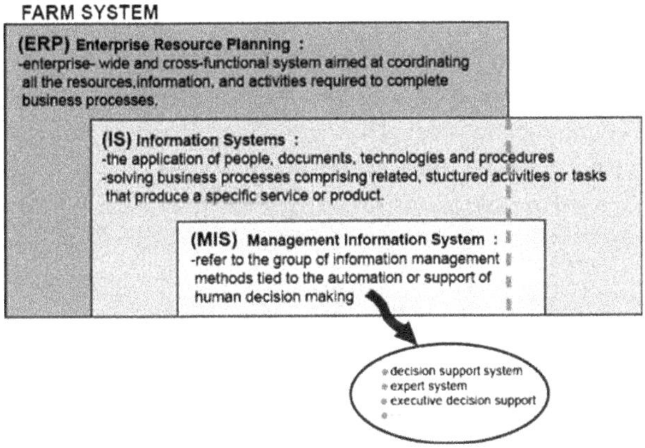

Figure 2 Concept of MIS (Sørensen et al., 2010, adapted from project FutureFarm report 3.1 System Analysis and Definition of System Boundaries).

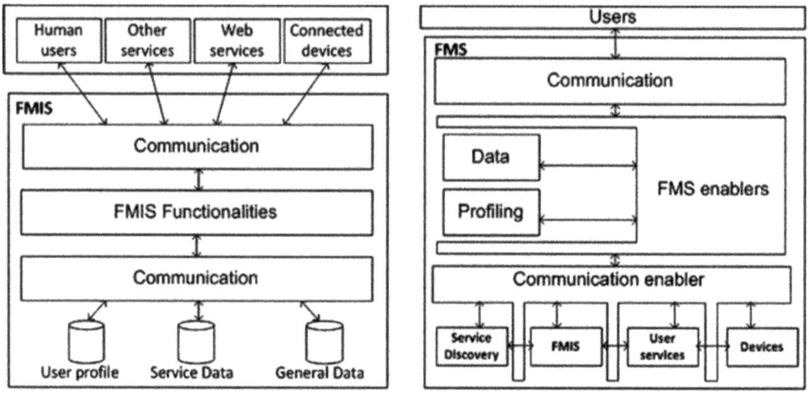

Figure 3 The difference between FMIS and FMS by Kaloxylos et al. (2012), where FMIS is a functional part of FMS.

system (AIMS) which is connected to national-level databases and may also include a submodule (potentially called an agriculture information and innovation system) utilizing big data applications. However, this approach potentially confuses an FMIS and an FMS (farm management system). Kaloxylos et al. (2012) have defined FMIS as part of a larger FMS as shown in Fig. 3.

3 Functional requirements and modular components of farm management information systems

Functionalities that are needed from a FMIS vary between the farms. The internal heterogeneity between farms, in such characteristics as location, farm size, production line(s), personnel and skills, economic status, land and animal type and properties, is vast. Similarly, external factors such as policy and regulations, market situation, climate, weather and environmental conditions, business partners, available financing, etc. vary from farm to farm. This means functionalities and applicability of information solutions are farm specific. However, a group of general-level functionalities common to most farms have been identified by researchers. Functionalities (based on needs of a farm manager to run farm operations (at MIS) level) identified by Sørensen et al. (2010) include:

- Monitoring and managing farm production processes;
- Managing machinery and tools;
- Managing human resources (workers);
- Managing the availability of production materials;
- Managing finances; and
- Taking care of external regulations or certification requirements.

Fountas et al. (2015) analysed and listed functionalities provided by commercial FMISs. The resulting functionalities and their descriptions are shown in Table 1.

Academic research is constantly searching for ways to improve the adoption of farm information management tools by identifying bottlenecks and developing generic solutions that commercial technology providers can adopt. However, the transition from research to practical applications takes time. As an example, Fountas et al. (2015) found that although researchers had prototyped and published new functionalities to enhance adoption of, e.g. precision agriculture, Internet of Things (IoT) technologies and cloud computing in FMIS solutions years earlier (Murakami et al., 2007; Pesonen et al., 2008; Nikkilä et al., 2010; Kaloxylos et al., 2012), the functionalities provided by commercial FMIS providers still lacked these features.

Different commercial FMISs typically provide a different set of functionalities depending on customer type and needs. For example, Fountas et al. (2015) identified 141 solutions for crop production in open fields and found the most frequently functions in FMIS included:

- Field operations management (63%);
- Reporting (e.g. on predicted and actual crop yields) (57%);
- Financial information (45%);
- Site-specific management (40%);
- Inventory management (38%);
- Machinery management (28%); and
- Human resource management (25%).

Less frequently used functions included:

- Traceability (19%);
- Quality assurance (19%);
- Sales (18%); and
- Good agricultural practices (16%).

Tummers et al. (2021) made a complete list of necessary functions which are summarized in Fig. 4.

Tummers et al. (2021) identified 29 key modules needed in an FMIS (Fig. 5). This provides a helpful checklist of components when studying or comparing different FMISs. In practice, the content of commercial FMISs varies and reflects the heterogeneity and complexity of farming and farm needs. Tummers et al. (2019), e.g., found a total of 81 separate features and 53 gaps in coverage in existing commercial FMISs.

Table 1 Eleven FMIS functions included in the commercial software summarised in a review by Fountas *et al* (2015)

Function title	Function description
Field operations management	Includes the recording of farm activities. This function also helps the farmer to optimize crop production by planning future activities and observing the actual execution of planned tasks. Furthermore, preventive measures may be initiated based on the monitored data
Best practice (including yield estimation)	Includes production tasks and methods related to applying best practices according to agricultural standards (e.g. organic standards, integrated crop management requirements). An yield estimate is feasible through the comparison of actual demands and alternative possibilities, given hypothetical scenarios of best practices
Finance	Includes the estimation of the cost of every farm activity, input-output calculations, labour requirements and so on, per unit area. Projected and actual costs are also compared and input into the final evaluation of the farm's economic viability
Inventory	Includes the monitoring and management of all production materials, equipment, chemicals, fertilizers, and seeding and planting materials. The quantities are adjusted according to the farmer's plans and customer orders. A traceability record is also an important feature of this function
Traceability	Includes crop recall, using an ID labelling system to control the produce of each production section. Traceability records related to the use of materials, employees and equipment can be easily archived for rapid recall.
Reporting	Generally includes the creation of farming reports, such as planning and management, work progress, work sheets and instructions, orders purchases cost reporting and plant information
Site specific	Includes the mapping of the features of the field. The analysis of the collected data can be used as a guide for applying inputs with variable rates. The goal of this function is to reduce or optimize input and increase output
Sales	Includes the management of orders, the packing management and accounting systems and the transfer of expenses between enterprises, charges for services, and the costing system for labour, supplies and equipment charge-outs
Machinery management	Includes the details of equipment usage, the average cost per work-hour or per unit area. It also includes fleet management and logistics
Human resource management	Includes employee management, including, e.g. the availability of employees in time and space. The goal is the rapid, structured handling of issues concerning employees, such as work times, payment, qualifications, training, performance and expertise
Quality assurance	Includes process monitoring and the production evaluation according to current legislative standards

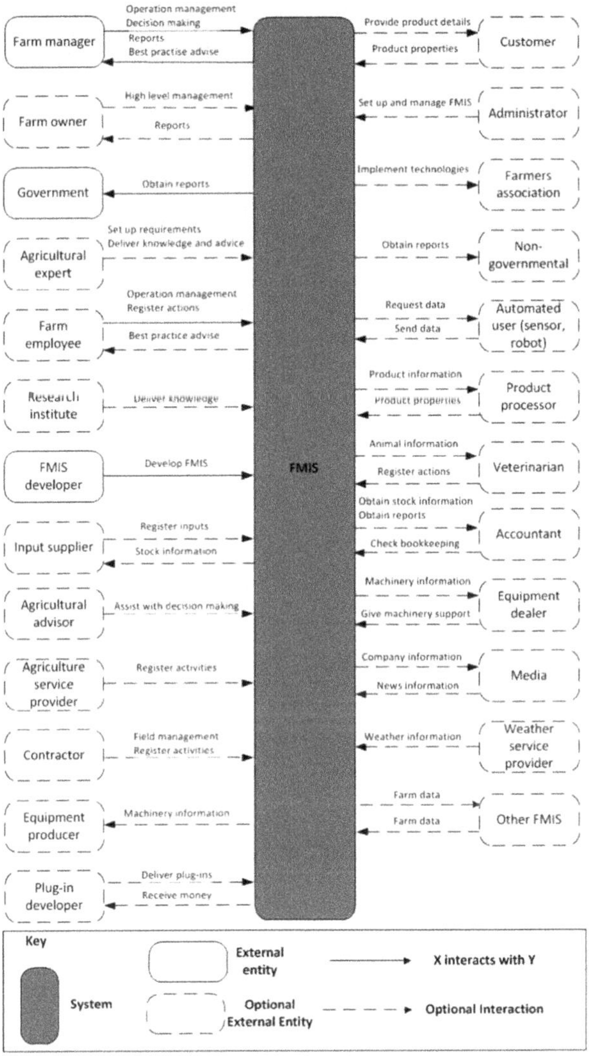

Figure 4 A context view of FMIS showing the overall purpose of the system and its interfaces with an external environment (Tummers et al., 2021).

4 The evolution of farm management information systems

Farm management has existed as long as we have had farms. Lewis (1998) describes the early development of FMIS in terms of differing levels of sophistication, depending on the technical tools adopted. The initial need for a system appeared with the need to report farm business activities, i.e. for

Figure 5 A decomposition view presenting different modules of FMIS (Tummers et al., 2021).

taxation purposes. Lewis named these early systems as types of farm record system (FRS), where the minimum requirement is to maintain a basic cash book. He grouped levels of sophistication as follows:

- No FRS, where cheque stubs and receipts are stored for use by an external accountant;
- Manual FRS where information provided is not used to support decision-making;

- Manual FRS where information from a FRS is used in farm business decision-making, possibly with the assistance of electronic calculators or basic office tools; and
- Computer-supported FMIS where farm businesses use a combination of computer and manual methods.

Personal computers (PC) became common in companies, households and farms in 1990s in tandem with the growth of the Internet (Batte et al., 1995; Batte, 2005; Lawson et al., 2011). This paved the way for commercial FMIS PC software development and adoption by farmers.

Early FMIS software typically provided agronomic expert knowledge which was hard coded in programs. This included recommendations for choosing crops and even varieties per field block depending on soil type, fertilizer recommendations per crop type, tracking inputs used as well as yields. The growth of government regulation and subsidy schemes added new requirements for financial and other information which was also hard coded into FMIS (Nikkilä et al., 2012).

There were, however, differences in availability and adoption rate of ICT technologies between the regions and different farming sectors (Holster et al., 2012). Early PC-based FMISs were typically developed to meet local needs and contained hard-coded local expertise such as local recommendations for use of farm inputs, usually allocated on a field-specific basis, based on sources such as local field trial data. Farmers also became familiar with other generic software products like spreadsheets or generic accounting solutions to support daily management (Batte et al., 1995).

When precision agriculture (PA), involving identifying in-field variation and adapting management accordingly, became more common in 2000s, FMIS faced new challenges. Management systems for task planning and yield monitoring in PA needed to be based on Geographic Information Systems (GIS), and mapping software was needed to identify separate management zones in fields for site-specific management. The greater complexity and cost of these technologies meant they were first developed by the larger multinational agricultural machine manufacturers who sold PA technology as add-on technology for use with equipment such as tractors. Over time, some of these solutions have grown into more all-encompassing FMIS solutions. This has led to a complex situation on some farms which may use equipment from different companies which all provide their own software solutions which vary from supporting practices such as controlled traffic farming through to more ambitious decision support tools for a wide range of tasks.

As these new technologies became more established, there were increasing calls for some kind of system integration. The 2010s saw the development of the so-called future internet (Gubbi et al., 2013) and Smart Farming or Agriculture 4.0 (corresponding to Industry 4.0, Liu et al., 2021; EU Commission, 2017) technologies. Other developments included the IoT, robotics, big data, artificial intelligence (AI), blockchain, cloud computing, next-generation mobile communication technologies, advanced sensor and sensor network technologies, increased calculation capacity, machine learning, service-oriented architecture (SOA), application programming interfaces (API) and Apps on mobile phones. These enabled the development of internet (or web) platforms capable of integrating different data sources and services. Research paved the way for FMISs that utilize Internet and IoT and SOA technologies (Murakami et al., 2007; Teye, 2011; Kaloxylos et al., 2012, 2014; Kruize et al., 2016; Köksal and Tekinerdogan, 2019). New system architecture models focus on developing a single FMIS using SOA and integrating it with latest sensing techniques, IoT and machinery technologies, analytics, decision support systems (DSS), etc.

FMISs have also evolved to utilize internet platforms which allow local FMISs and global solutions to collaborate to form a modular farm management system (FMS) of systems with data connections which better serve farms' needs (Kaloxylos et al., 2012). Internet platform-based FMIS such as SaaS (software as a service) are now gaining ground from local FMISs (PC software), not only due to the improved integration possibilities for different services but also due to access to information and services such as task planning, available at any time through mobile internet connections.

The concept of internet platform-based FMIS has also enabled low-cost solutions that use mobile Apps as a user interface providing agronomic expertise for dedicated farming tasks. In these solutions, the farmer usually inputs farm-specific data manually while the cloud backend utilizes open data sources to provide updated information (Henriyadi et al., 2022). An example of a commercial solution following this principle is Yara's AtFarm service (https://www.at.farm/). This provides AI-aided fertilization planning assistance based on satellite images and farmers manual inputs via a mobile App. Another example is Amazon Retail which provides AI-assisted agronomy services for farmers in India (https://www.aboutamazon.in/news/amazon-india-news/amazon-retail-launches-agronomy-services-to-empower-farmers).

There has now been continuous technical development of FMIS over decades. For example, Schulze Schwering et al. (2022), list 13 commercial FMISs used globally. However, adoption of FMISs still varies significantly, depending on farm size, crops or livestock produced, economic status and business models, as well as how well an FMIS fits the needs of a particular farm and how comfortable a farmer is with using digital tools (Schulze Schwering et al., 2022). Schulze Schwering et al. studied the adoption of FMISs among 280 German farmers in 2019 and identified 58% as 'users'. This suggests that much more work is needed to ensure FMISs meet farmers' needs and to incentivize and enable farmers to use them.

5 Reference software architectures for farm management information systems

Integrating different digital solutions in an FMIS requires collaboration between those developing FMIS and other solution providers. This has created the need to share reference architectures of FMIS, e.g. functional architectures showing the functional components of the system. Reference software architectures are also shared which helps FMIS developers. Figure 6 illustrates a generic platform architecture of a software ecosystem used in the FIspace EU project (Kruize et al., 2016). The architecture consists of several separate functional software components such as system and data integration, operating system, security and privacy safeguarding systems, orchestration (e.g. to coordinate separate tasks into an overall workflow), etc.

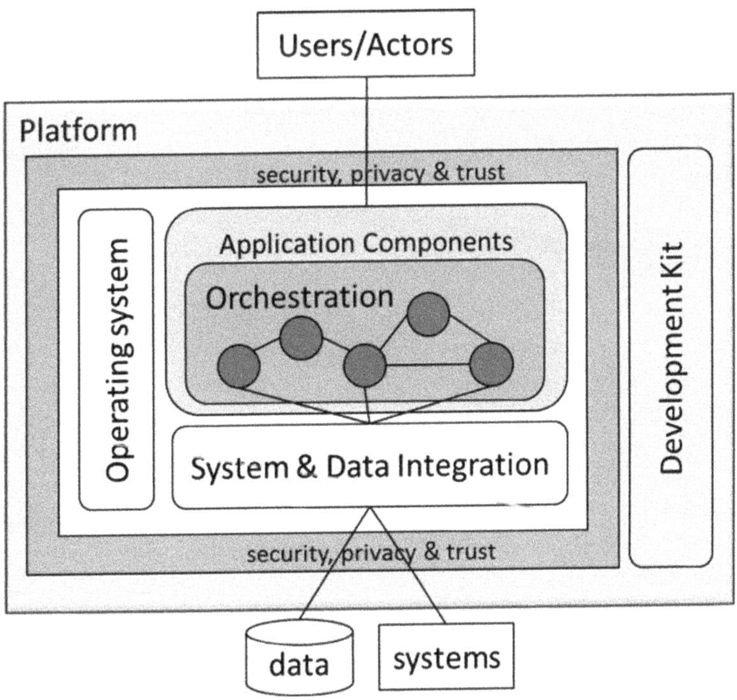

Figure 6 Basic architecture of a platform for a farm software ecosystem. Adapted from the FIspace platform (Kruize et al., 2016).

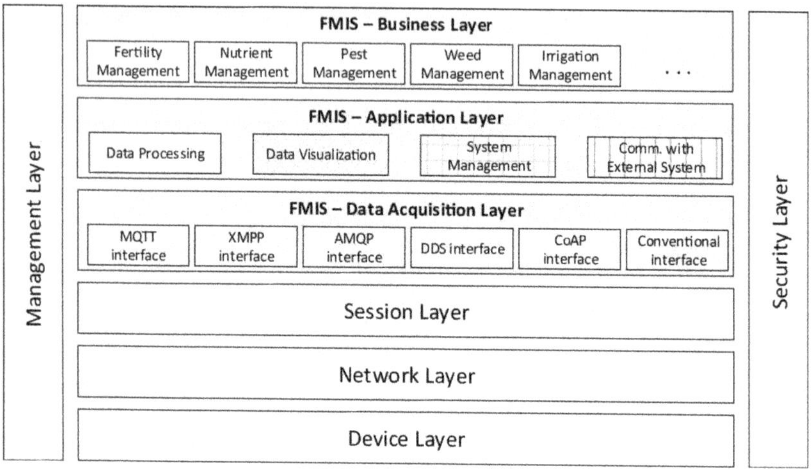

Figure 7 IoT-based FMIS: layered view (Köksal and Tekinerdogan, 2019).

There are several ways of visualizing a system architecture:

- A decomposition approach to identify separate functional components (Fig. 5);
- A layered approach where the hierarchy of layers identifies software modules and the order in which they communicate between layers (Fig. 6); and
- A deployment approach which shows the allocation of software modules to the hardware of a computing platform (Köksal and Tekinerdogan, 2019).

Figure 7 shows an architecture design approach for an IoT-based FMIS, where a device layer is an integral part of the FMIS.

One of the reference architectures paving the way towards a tailorable FMIS as a system of systems has been developed by the FIWARE community (FIWARE, 2018, Fig. 8). The architecture is based on modular open-source solutions which enables building a comprehensive farm management system. The FMIS focusing on farm operation support is freely tailorable within the architecture but is only a part of FMS. This type of architecture makes it possible to tailor solutions to a specific (often small) user group. Solutions like AgroDSS (Rupnik et al., 2019), for example, could be deployed in these kinds of system.

6 Conclusion and future trends

There has been a long period of evolution in the development of FMIS from simple, single FRS through isolated PC software to ecosystem-type system of systems technologies functioning in the cloud. The sophistication and

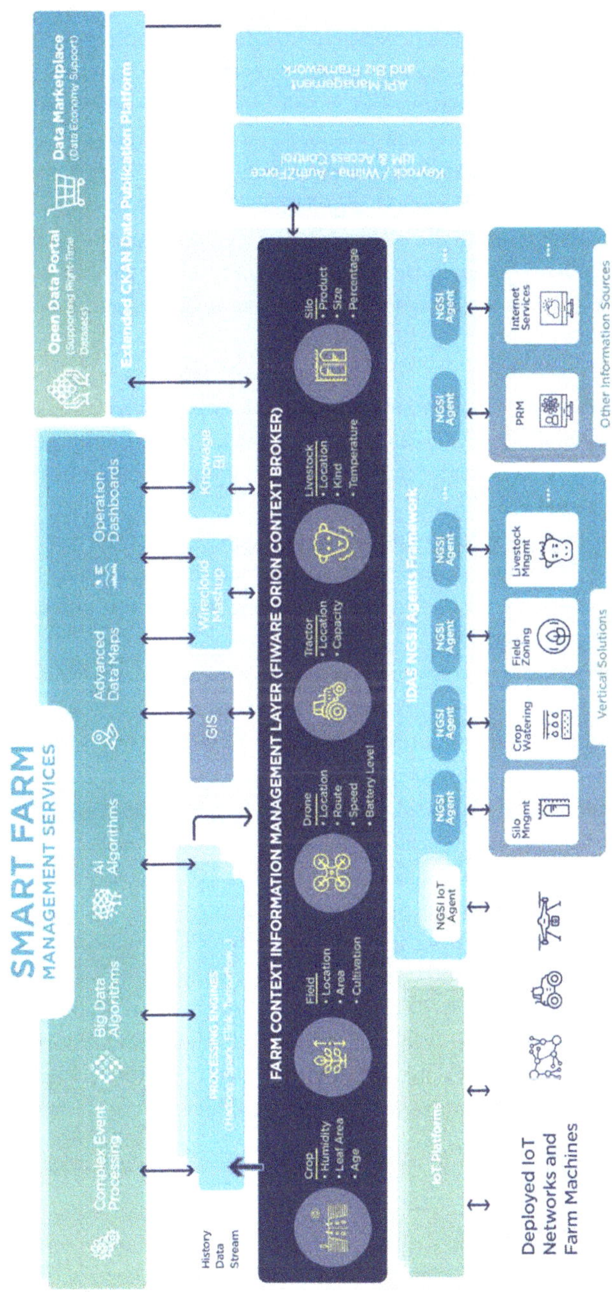

Figure 8 A reference architecture for smart farm management system powered by FIWARE (FIWARE, 2018).

functionalities of systems have evolved in response to wider developments in information technologies and the changing requirements of farms for information (Köksal and Tekinerdogan, 2019). The adoption rate of FMISs in farms, which varies from region to region and from one type of production to another, is slowly increasing but is still low. Today's open source modular FMIS prototypes, utilizing service-oriented architecture and Agriculture 4.0 and novel internet technologies, offer the potential for individually customable FMIS solutions for each farm. However, significant FMIS development is needed in the future in order to meet the needs of and to attract the interest of all farmers. The transition from research prototypes to commercial products needs to be faster than it is currently. Schulze Schwering et al. (2022) propose seven areas for optimization of FMIS:

- Further development and investment in compatibility;
- More user-friendly FMIS (e.g. with more intuitive user interfaces);
- More activity to promote awareness (and the practical benefits) of FMIS;
- Building trust in FMIS and data security;
- Ensuring sufficient and reliable internet coverage;
- Aligning FMIS with common documentation requirements in order to allow farmers to digitally submit relevant data to the authorities from a single source; and
- More trainings for farmers on how to implement and use FMIS successfully.

Looking to the future in Europe, the EU's data strategy aims to enhance digitalization throughout society (European Commission, 2020). The aim is to build a fair European data economy respecting the data sovereignty of individuals and organisations and right of those producing data to benefit from their data, including farmers. So-called data space structures with controlled consent management by farmers for data sharing and utilization will affect the composition of FMISs, especially data handling, storing and analytics functionalities. Farmers will have new needs and rights to use their data that will need to be reflected in FMIS structures. New types of data intermediation services will enable digital actors to create agile data connections and enable data flows within and between sectors and across borders. One result could be development of ecosystem-type farm-specific FMISs. There are several initiatives to create a supporting legal framework for such a transition. Among the most important are:

- Data Governance Act;
- Data Act;
- AI Act;
- Digital Market Act; and

- Digital Services Act.

A brief introduction to this topic in given by Sitra (2022). The Data Spaces Business Alliance (DSBA, https://internationaldataspaces.org/wp-content/uploads/dlm_uploads/Data-Spaces-Business-Alliance-Technical-Convergence.pdf/) has brought several organisations to work together to standardize architecture components as well as to discuss legal, ethical and policy issues. The EU-funded DEMETER Project is also contributing to this development with its reference architecture and new system components (Roussaki et al., 2022). The new components are based on knowledge sharing and co-creation by technology providers to ensure interoperability and a common dashboard for all stakeholders.

Global actors are also entering the farm information management sector, providing agricultural knowledge to farmers based on open data sources and AI-assisted data analysis, as well as mobile Apps functioning as User Interfaces (UI). Typically, services are provided to farmers as low-cost services with some farm data already provided (Yara AtFarm, Amazone Retail agronomic services). The sector is open also for local technology and business providers to develop solutions, i.e. utilizing open-source components to develop more attractive solutions for local farmers to adopt.

7 Where to look for further information

A book chapter by Tsiropoulos et al. (2017) gives a nice overview of diverse aspects of FMIS including functionalities and applications, cost structures, user adoption, cost structures and adoption.

- Tsiropoulos, Z., Carli, G., Pignatti, E. and Fountas, S. (2017). Future perspectives of Farm Management Information Systems. In S. M. Peredsen and K. M. Lind (Eds.), *Precision Agriculture: Technology and Economic Perspectives* (pp. 181-200). Cham: Springer. https://ebookcentral.proquest.com/lib/nrifi-ebooks/reader.action?docID=5149553.

Tummers et al. (2019) gives a comprehensive understanding of present status and development need of FMISs.

- Tummers, J., Kassahun, A. and Tekinerdogan, B. (2019). Obstacles and features of Farm Management Information Systems: A systematic literature review. *Computers and Electronics in Agriculture* 157,189-204. https://doi.org/10.1016/j.compag.2018.12.044.

Roussaki et al. (2022) explains the recent advances in interoperability of systems when building FMIS as a system of system.

- Roussaki, I., Doolin, K., Skarmeta, A., Routis, G., Lopez-Morales, J. A., Claffey, E., Mora, M. and Martinez, J. A. (2022). Building an interoperable space for smart agriculture. Digital Communications and Networks (in press). https://doi.org/10.1016/j.dcan.2022.02.004.

Schulze Schwering et al. (2022) gives a comprehensive insight to reasons why adoption rates of FMISs and digital tools in general are low.

- Schulze Schwering, D., Bergmann, L. and Sonntag, W. I. (2022). How to encourage farmers to digitize? A study on user typologies and motivations of farm management information systems. *Computers and Electronics in Agriculture* 199, 107133. https://doi.org/10.1016/j.compag.2022.107133.

8 References

Batte, M. T. (2005). Changing computer use in agriculture: Evidence from Ohio. *Computers and Electronics in Agriculture* 47(1), 1–13. https://doi.org/10.1016/j.compag.2004.08.002.

Batte, M. T., Rister, M. E., Frank, G. and Schnitkey, G. D.(1995). Farm information systems design and use. In Battle, M. T. *Adoption and Use of Farm Information Systems*. North Central Regional. Research Publication 339, OARDC Special Circular 149, November (Chapter 2). Available at: https://kb.osu.edu/handle/1811/71890 (accessed 22 October 2022).

Burlacu, G., Costa, R., Sarraipa, J., Jardim-Golcaves, R. and Popescu, D. (2014). A conceptual model of farm management information system for decision support. 5th Doctoral Conference on Computing, Electrical and Industrial Systems (DoCEIS), April 2014, Costa de Caparica, pp. 47–54. https://doi.org/10.1007/978-3-642-54734-8_6.

European Commission (2017). Industry 4.0 in agriculture: Focus on IoT aspects. Digital Transformation Monitor. Internal Market, Industry, Entrepreneurship and SMEs. Available at: https://ati.ec.europa.eu/reports/technology-watch/industry-40-agriculture-focus-iot-aspects (accessed 23 October 2022).

European Commission (2020). Shaping Europe's digital future. Publications Office of the European Union, Luxembourg. ISBN 978-92-76-16363-3 https://doi.org/10.2759/091014.

FIWARE (2018). The open source platform of choice for building smart farming solutions. Available at: https://www.fiware.org/wp-content/uploads/2018/06/Smart-Agrifood_Brochure_June2018.pdf.

Fountas, S., Carli, G., Sørensen, C. G., Tsiropoulos, Z., Cavalaris, C., Vatsanidou, A., Liakos, B., Canavari, M., Wiebensohn, J. and Tisserye, B. (2015). Farm management information systems: Current situation and future perspectives. *Computers and Electronics in Agriculture* 115, 40–50. https://doi.org/10.1016/j.compag.2015.05.011.

Gubbi, J., Buyya, R., Marusic, S. and Palaniswami, M. (2013). Internet of Things (IoT): A vision, architectural elements, and future directions. *Future Generation Computer Systems* 29(7), 1645–1660. https://doi.org/10.1016/j.future.2013.01.010.

Henriyadi, H., Esichaikul, V. and Anutariya, C. (2022). A conceptual model for development of small farm management information system: A case of Indonesian smallholder chili farmers. *Agriculture* 12(6), 866. https://doi.org/10.3390/agriculture12060866.

Holster, H., Horakova, S., Ipema, B., Fusai, B., Teye, F., Martini, D., Shalloo, L. and Schmid, O. (2012). Current situation on data exchange in agriculture in the EU27 & Switzerland. D2.4 Final Report Project AgriXchange - Network for Data Exchange in Agriculture. Available at: https://edepot.wur.nl/206268 (accessed 24 October 2022).

Kaloxylos, A., Eigenmann, R., Teye, F., Politopoulou, Z., Wolfert, S., Shrank, C., Dillinger, M., Lampropoulou, I., Antoniou, E., Pesonen, L., Nicole, H., Thomas, F., Alonistioti, N. and Kormentzas, G. (2012). Farm management systems and the Future Internet era. *Computers and Electronics in Agriculture* 89, 130–144. https://doi.org/10.1016/j.compag.2012.09.002.

Kaloxylos, A., Groumas, A., Sarris, V., Katsikas, L., Magdalinos, P., Antoniou, E., Politopoulou, Z., Wolfert, S., Brewster, C., Eigenmann, R. and Maestre Terol, C. (2014). A cloud-based farm management system: Architecture and implementation. *Computers and Electronics in Agriculture* 100, 168–179. https://doi.org/10.1016/j.compag.2013.11.014.

Köksal, Ö. and Tekinerdogan, B. (2019). Architecture design approach for IoTbased farm management information systems. *Precision Agriculture* 20(5), 926-958. https://doi.org/10.1007/s11119-018-09624-8.

Kruize, J. W., Wolfert, J., Scholten, H., Verdouw, C. N., Kassahun, A. and Beulens, A. J. M. (2016). A reference architecture for Farm Software Ecosystems. *Computers and Electronics in Agriculture* 125, 12–28. https://doi.org/10.1016/j.compag.2016.04.011.

Lawson, L. G., Pedersen, S. M., Sørensen, C. G., Pesonen, L., Fountas, S., Werner, A., Oudshoorn, F. W., Herold, L., Chatzinikos, T., Kirketerp, I. M. and Blackmore, S. (2011). A four nation survey of farm information management and advanced farming systems: A descriptive analysis of survey responses. *Computers and Electronics in Agriculture* 77(1), 7-20. https://doi.org/10.1016/j.compag.2011.03.002.

Lewis, T. (1998). Evolution of farm management information systems. *Computers and Electronics in Agriculture* 19(3), 233–248. https://doi.org/10.1016/S0168-1699(97)00040-9.

Liu, Y., Ma, X., Shu, L., Hancke, G. P. and Abu-Mahfouz, A. M. (2021). From Industry 4.0 to Agriculture 4.0: Current status, enabling technologies, and research challenges. *IEEE Transactions on Industrial Informatics* 17(6), 4322–4334. https://doi.org/10.1109/TII.2020.3003910.

Murakami, E., Saraiva, A. M., Ribeiro Jr., L. C. M., Cugnasca, C. E., Hirakawa, A. R. and Correa, P. L. P. (2007). An infrastructure for the development of distributed service oriented information systems for precision agriculture. *Computers and Electronics in Agriculture* 58(1), 37–48. https://doi.org/10.1016/j.compag.2006.12.010.

Nikkilä, R., Seilonen, I. and Koskinen, K. (2010). Software architecture for farm management information systems in precision agriculture. *Computers and Electronics in Agriculture* 70(2), 328–336. https://doi.org/10.1016/j.compag.2009.08.013.

Nikkilä, R., Wiebenschon, J., Nash, E., Seilonen, I. and Koskinen, K. (2012). A service infrastructure for the representation, discovery, distribution and evaluation of agricultural production standards for automated compliance control. *Computers and Electronics in Agriculture* 80, 80–88. https://doi.org/10.1016/j.compag.2011.10.011.

Paraforos, D. S., Vassiliadis, V., Kortenbruck, D., Stamkopoulos, K., Ziogas, V., Sapounas, A. A. and Griepentrog, H. W. (2016). A farm management information system using future internet technologies. *IFAC-PapersOnLine* 49(16), 324–329. https://doi.org/10.1016/j.ifacol.2016.10.060.

Pesonen, L., Koskinen, H. and Rydberg, A. (2008). InfoXT - user-centric mobile information management in automated plant production. Nordic Innovation Centre, Finland. Available at: https://norden.diva-portal.org/smash/record.jsf?pid=diva2%3A707134&dswid=-4155.

Roussaki, I., Doolin, K., Skarmeta, A., Routis, G., Lopez-Morales, J. A., Claffey, E., Mora, M. and Martinez, J. A. (2022). Building an interoperable space for smart agriculture. *Digital Communications and Networks*, 9 (1), 183-193. https://doi.org/10.1016/j.dcan.2022.02.004.

Rupnik, R., Kukar, M., Vračar, P., Košir, D., Pevec, D. and Bosnić, Z. (2019). AgroDSS: A decision support system for agriculture and farming. *Computers and Electronics in Agriculture* 161, 260–271. https://doi.org/10.1016/j.compag.2018.04.001.

Schulze Schwering, D., Bergmann, L. and Sonntag, W. I. (2022). How to encourage farmers to digitize? A study on user typologies and motivations of farm management information systems. *Computers and Electronics in Agriculture* 199, 107133. https://doi.org/10.1016/j.compag.2022.107133.

Sitra (2022). EU regulation builds a fair data economy. The Opportunities of the Big Five Proposals for Businesses, Individuals and the Public Sector. Working paper. Available at: https://www.sitra.fi/app/uploads/2022/06/sitra-eu-regulation-builds-a-fairer-data-economy.pdf.

Sørensen, C. G., Fountas, S., Nash, E., Pesonen, L., Bochtis, D., Pedersen, S. M., Basso, B. and Blackmore, S. B. (2010). Conceptual model of a future farm management information system. *Computers and Electronics in Agriculture* 72(1), 37–47. https://doi.org/10.1016/j.compag.2010.02.003.

Teye, F. (2011). A conceptual model for collaboration-based management information systems. Master Thesis, Metropolia University of Applied Science, Helsinki. Available at: https://urn.fi/URN:NBN:fi:amk-201105096954.

Tsiropoulos, Z., Carli, G., Pignatti, E. and Fountas, S. (2017). Future perspectives of Farm Management Information Systems. In S. M Peredsen and K. M. Lind (Eds.) *Precision Agriculture: Technology and Economic Perspectives* (pp. 181–200). Cham: Springer. Available at: https://ebookcentral.proquest.com/lib/nrifi-ebooks/reader.action?docID=5149553.

Tummers, J., Kassahun, A. and Tekinerdogan, B. (2019). Obstacles and features of Farm Management Information Systems: A systematic literature review. *Computers and Electronics in Agriculture* 157, 189–204. https://doi.org/10.1016/j.compag.2018.12.044.

Tummers, J., Kassahun, A. and Tekinerdogan, B. (2021). Reference architecture design for farm management information systems: A multicase study approach. *Precision Agriculture* 22(1), 22–50. https://doi.org/10.1007/s11119-020-09728-0.

Van der Ploeg, J. D. (2016). Family farming in Europe and Central Asia: History, characteristics, threats and potential. International Policy Centre for Inclusive Growth (IPC-IG) Working Paper No. 1. ISBN: 978-92-5-109538-6 (FAO). Available at: http://www.fao.org/3/a-i6536e.pdf.

Zysk, E., Mroczkowski, T. and Dawidowicz, A. (2022). Small farms as "data producers" for the needs of agricultural management information system. *Geomatics and Environmental Engineering* 16(3), 79–109. https://doi.org/10.7494/geom.2022.16.3.79.

Chapter 2

The role of digital technologies in achieving sustainable agriculture

Thiago L. Romanelli, André F. Colaço and João P. S. Veiga, University of São Paulo, Brazil

1 Introduction

Human history can be divided into different periods or ages, depending on developments such as innovations in technology and organization interacting with variables such as available land and resources to exploit (Fig. 1). In terms of agriculture and food production, the first of these periods is the Hunter-Gatherer Age characterized by nomadic tribes moving from place to place to exploit the available seasonal resources, whether hunting wild game or gathering plants (Hamilton et al., 2016; Naylor, 1996; Hémery et al., 1993).

Although hominids started to plant food and herd animals as early as 80 000 BC, it was only in the Neolithic period that farming developed on a significant scale, thus starting a second major phase of development: the Agricultural Age. The lowlands along the Nile, Tigris, and Euphrates rivers, where seasonal flooding enriched soils, formed the so-called Fertile Crescent. These allowed Egypt and Mesopotamia (the area now occupied by Iraq) to dramatically increase the production of cereals and other plants. The resulting food surplus allowed both a significant population increase and larger, more complex societies, including greater division of labor to develop a range of other skills (e.g. in writing, pottery, and metal working) which then further contributed to improving food production (Tudge, 1998; Hémery et al., 1993).

Despite its many achievements, production in the Agricultural Age remained fundamentally constrained by reliance on organic inputs, the limits

http://dx.doi.org/10.19103/AS.2023.0132.10

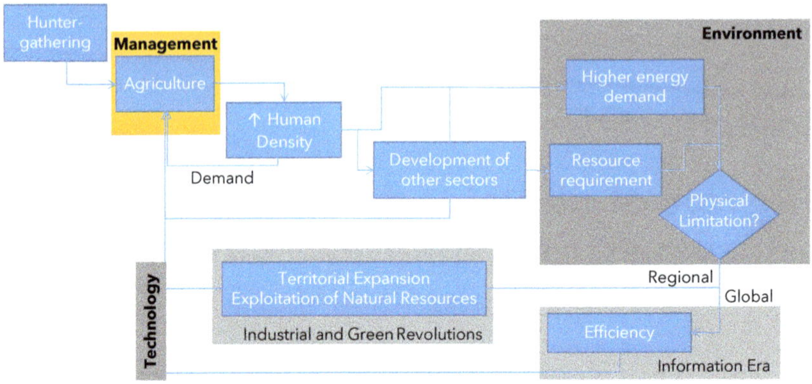

Figure 1 Agricultural-human development as determined by resource availability and scale.

imposed by a primary reliance on humans and animals as a source of power for activities such as ploughing, planting and harvesting, and vulnerability to abiotic stresses such as poor weather and biotic threats such as pests and diseases. These constraints started to be addressed by the third age of human development: the Industrial Age. Major developments in technology meant new forms of power (starting with the steam engine) that could dramatically increase the scale and efficiency of production (Tudge, 1998; Passet, 1984). At the same time, major advances in science laid the foundations for the first synthetic fertilizers and pesticides as well as modern, more precise breeding techniques. This period was also characterized by a significant expansion in trade and a wave of colonization of other countries by the Western nations which laid the foundations for a major global expansion in agriculture and a globalized system of food production and supply. Some of these developments can be seen, e.g. in the green revolution of the 1960s in which a combination of modern crop varieties and a new generation of fertilizers and pesticides allowed dramatic increases in agricultural production in many developing countries to feed rapidly growing populations (Tilman et al., 2002; Cassman, 1999).

However, we are now seeing the limits of this Industrial Age in many sectors, including agriculture. The huge input demands of modern agriculture, in terms of raw materials needed to manufacture the resources on which it depends and the energy (primarily from fossil fuels) required to power farming machinery and infrastructure, are threatening to exhaust the world's available natural resources. At the same time, the global expansion of agricultural production is becoming environmentally unsustainable, e.g. with rates of deforestation contributing to potentially catastrophic climate change and biodiversity loss. Emissions from modern, intensive agricultural production, including nutrient losses through leaching of synthetic fertilizers, are polluting the environment and contributing further to climate change. Agriculture seems to be reaching limits allowed by fundamental processes such as photosynthesis, evapotranspiration, and biomass production (Tilman et al., 2002; Waggoner, 1995).

These resource and territorial limits mean that to become both more sustainable and increase output to meet the needs of continued global population growth, agriculture must now focus on becoming much more efficient in its use of resources (Bentley, 2013; Bois, 1985). This coincides with the huge increases since the Second World War in information gathering, communication, and processing technologies, driven in part by developments in computing, which constitute a fourth phase in human development: the Information or Knowledge Age. These advances in the ability to gather and analyze large amounts of data offer an opportunity to better track, analyze, and optimize resource use in sectors such as agriculture. As Marchetti (2009) states in his analysis of energy ratios in food production, solutions need to be focused more on ingenuity in using available technology more efficiently rather than on just developing more technologies in such areas as energy production, i.e. 'more bits and less kilowatts'.

Recent advances in technology have been characterized using terms such as smart sensors, internet of things (IoT), cloud computing, big data, artificial intelligence (AI), and machine learning. However, there is a risk of focusing too much on the available tools and what they can do rather than on the problems and solutions they can be used to address. As the comment by Marcheti suggests, information is only useful if problems are clearly understood and framed so that the appropriate data can be gathered, processed, and analyzed to inform solutions able to deliver real improvements on the ground.

2 Efficiency indicators in agriculture

Modern agriculture has enabled the production of food, fiber, and plant-based energy (e.g. biofuels) on an unprecedent scale. However, some indicators cast doubt on its ability to produce these resources in an efficient manner. Natural resources for production may be underutilized. As an example, the Global Yield Gap and Water Productivity Atlas (2022) suggest that the production of corn, soybean, and sugarcane in Brazil is around 57% of what could be achieved under optimal agronomic management. Similar numbers are found for wheat in USA and Australia. The use of agricultural inputs may also be inefficient. In the case of nitrogen fertilization for sugarcane in Brazil, Franco et al. (2008) found that only 20–30% of the applied nutrient was actually used by the crop, despite the fact that N fertilization represents around 20% of the total production cost.

Such inefficiencies are partially due to the nonlinear relationship between the amount of inputs used in crop production and the yield produced as output. In other words, for any increase in the use of inputs, the increase in the output production is invariably much lower. For example, the comparative growth in consumption of agricultural inputs and in the productivity of the main crops grown in Brazil since the Green Revolution show significant differences (Table 1). In terms of energy, this increase in the adoption of inputs to meet production demands means a greater amount of energy – supplied mainly by the use of fossil energy – has been incorporated into production systems in the form of fertilizers, pesticides, machinery, and equipment (Romanelli, 2021).

Overall, there is a need to improve agricultural production efficiency from a holistic perspective, including consideration of efficiency in energy consumption (and energy transformation) to minimize negative impacts on the environment, since, to a large extent, the use of fossil fuels is a major factor in greenhouse gas (GHG) emissions, in addition to representing a significant part of the cost of production. Agricultural systems are economic, thermodynamic, and physical entities, simultaneously subject to constraints in all these areas, demanding a systemic approach to the planning and evaluation of agricultural operations.

The data in Table 1 confirm that the Green Revolution greatly increased crop production from the middle of twentieth century to the present, including tripling yields of staple crops like rice and maize, but yield increases are now slowing or even stalling (Rong et al., 2021) despite growing consumption of fertilizers. The effort to keep reducing yield gaps through increased use of fertilizers can be counter-productive. In the majority of cases, application efficiency of fertilizers is below 50% (Rashmi et al., 2020), which means that

Table 1 Input consumption and yield from green revolution in Brazil

Input	Consumption variation (%)	Period
N	7829.6	1961-2016
P_2O_5	3349.5	1961-2016
K_2O	7999.3	1961-2016
Insecticide	289.7	1990-2015
Herbicide	861.4	1990-2015
Fungicide	685.9	1990-2015
Tractors	994.5	1961-2006
Crop land	87.7	1961-2015
Crop	**Yield variation (%)**	**Period**
Beans	49.7	1961-2016
Maize	226.8	1961-2016
Rice	221.6	1961-2016
Soybeans	157.8	1961-2016
Sugarcane	73.0	1961-2016

Source: Adapted from: Romanelli (2021) and FAO (2017).

much of the increase ends up contaminating water bodies and causing eutrophication as well as contributing to GHG emissions.

Different techniques can be used in the attempt to reduce fertilizer losses like coating the fertilizers with synthetic (e.g. polystyrene and polyethylene) or natural (e.g. bentonite, cellulose, starch, and biochar) polymers (Dong et al., 2020). In addition, a study in the south of Brazil found that 44% of the increase in soybean yields came from improvements in management with use of the appropriate variety and maturity group, the right plant density, and sowing date (Minussi Winck et al., 2023). Approaches to reduce loss and waste of food are also essential to make better use of the resources already invested (Lipinski et al., 2017).

3 From precision to digital agriculture

Agricultural science is constantly seeking to overcome inefficiencies in production by improving knowledge of processes and creating new technologies which, in turn, lead to improved, more efficient agronomic practices. Precision Agriculture (PA), Digital Agriculture (DA), and more recently Intelligent (or 'Decision') Agriculture are good examples of such advances. These technologies are often mistakenly referred to as if they were the same. However, it is important to understand the different concepts behind them and the way they represent important movements in recent technological development.

PA is a concept developed during the 1990s. According to the International Society of Precision Agriculture (ISPA), PA can be defined as 'a management strategy that gathers, processes and analyses temporal, spatial and individual data and combines it with other information to support management decisions according to estimated variability for improved resource use efficiency, productivity, quality, profitability and sustainability of agricultural production'. Other definitions are listed by the Research Group in AgroICT & Precision Agriculture at the Universitat de Lleida, Spain (Anonymous, 2022). A good recent overview of PA technologies is provided by Stafford (2019). The basis of PA lies in understanding spatial and temporal variability of crop and soil characteristics within fields. This understanding of in-field spatial and temporal variability is used to adapt management to meet the specific requirements of different sites within a field in contrast to uniform field-wide applications of inputs such as fertilizers and pesticides found in non-PA systems. This type of management is known as 'site-specific management' and it is made possible by technologies such as proximal and remote sensors, satellite positioning systems, and variable-rate application technologies. Key technological milestones in the development of PA have included the first-yield monitors in the late 1980s (Schueller and Bae, 1987) and the availability of GPS (global positioning systems) for civil use in the early 2000s.

Since the early developments of PA, numerous studies have reported improvements in resource use efficiency from site-specific management. Normally this outcome would have been achieved either by reducing input levels while maintaining crop yields (inputs used more rationally and effectively) or by increasing input use to achieve higher crop yields (by boosting areas of higher yield potential in the field). The early reviews of Lambert and Lowenberg-deBoer (2000) and Bongiovanni and Lowenberg-deBoer (2004) present several examples of the use of variable rate technology for the management of nutrients, weeds, pests, and water with varying positive impacts on the profitability and environmental protection of production systems. More recently, Colaço and Bramley (2018) reviewed the outcome of sensor-based technology for nitrogen (N) management in cereal crops, one of the most studied topics in precision agriculture research (Misara et al., 2022). Results show that sensor-based N application provided fertilizer savings ranging between 5% and 45% without significant impact on grain yield and an average profit increase of around US\$30 ha^{-1}.

The early use of digital technologies at the farm level made PA the precursor of DA. This is a more recent, broader, and more overarching concept than PA, referring to a broad range of applications of 'digital technologies' in agriculture (Fig. 2). While DA may have developed from PA and may employ similar technology, Shepherd et al. (2020) define DA more broadly as the 'ability to utilise the technology to convert precise data along the value chain into actionable knowledge to drive and support complex decision making on-farm and along the value chain'. Other discussions of DA use similar definitions, e.g. Cook et al. (2021). In general, DA can be seen as the wide-ranging encounter between agriculture and information and communication technology (ICT). This encounter has been characterized by some as a fourth agricultural revolution (or 'Agriculture 4.0') (Rose and Chilvers, 2018).

It is important to note that PA and DA do not entirely coincide (Fig. 2). DA moves beyond site-specific management characteristic of PA to cover all aspects not just of farming operations but those along the whole food supply chain. Likewise, not all PA practices are necessarily based on digital technologies. The understanding of spatial variability and site-specific management practices typical of PA systems predate, to some degree, the advent of digital technologies; for that reason, some may argue that the PA 'philosophy' is as old as agriculture itself. For example, farmers have long used traditional techniques such as visual inspection to identify spatial differences in crop health or growth to adapt agronomic practices accordingly. As such, this type of practice applies PA principles without the need for any specific DA technology. The role of modern PA (based on digital technologies) has been to help farmers identify

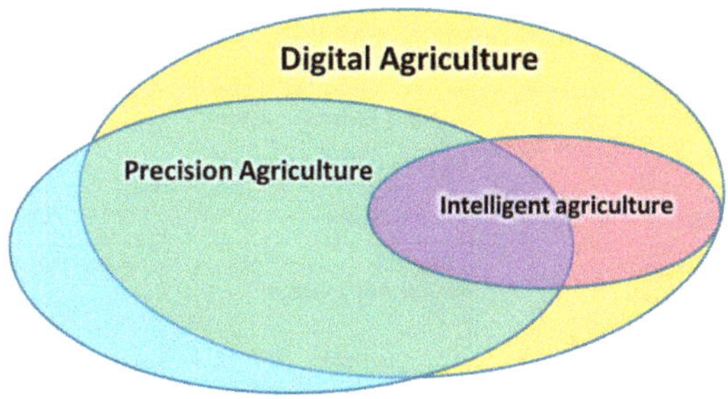

Figure 2 Digital Agriculture (Agriculture + ICT) encompassing Precision Agriculture and Intelligence Agriculture (Agriculture + AI).

spatial and temporal variability far more accurately and systematically at greater scales and to give them better tools to target input application more precisely and efficiently.

Another area of increasing attention in DA has been the use of information technology toward enabling 'complex decision-making on-farm and along the value chain' (Shepherd et al., 2020). This field of DA reflects, more specifically, the use of machine learning and other fields of artificial intelligence (AI) for decision-making in agriculture; in other words, it is the use of modern statistical methods and powerful software capabilities to transform digital data into actionable on-farm decisions. This movement has been often referred to as 'Intelligent Agriculture'. Analogous to the definition of DA (i.e. agriculture + ICT), 'Intelligent Agriculture' can be seen as the encounter between AI (or 'data science') and agriculture. Many examples can be found in Liakos et al. (2018), Chlingaryan et al. (2018), Klompenburg et al. (2020), and Colaço et al. (2021).

The development and potential of DA have been enabled by a huge expansion in the infrastructure needed to support it. According to the Word Bank (2022), Internet access increased from 7% to approximately 50% of the world population between 2000 and 2018, reaching 60% in 2020. This has supported developments such as cloud computing, the IoT, and mobile applications. There has also been a large growth in networks of satellites able to support global positioning satellite systems (GNSSs) and geographic information systems (GISs). The range and sophistication of sensor technologies able to measure crop and soil characteristics has also increased dramatically, as have the range of technologies on which they can be mounted, including proximal remote sensors (e.g. mounted on field machinery) and remote sensors mounted on unmanned aerial vehicles (UAVs) or satellites. Developments in AI have made it easier to analyze the exponential growth in available data.

In general, on-farm digital technologies can be classified into three main types (Fig. 3):

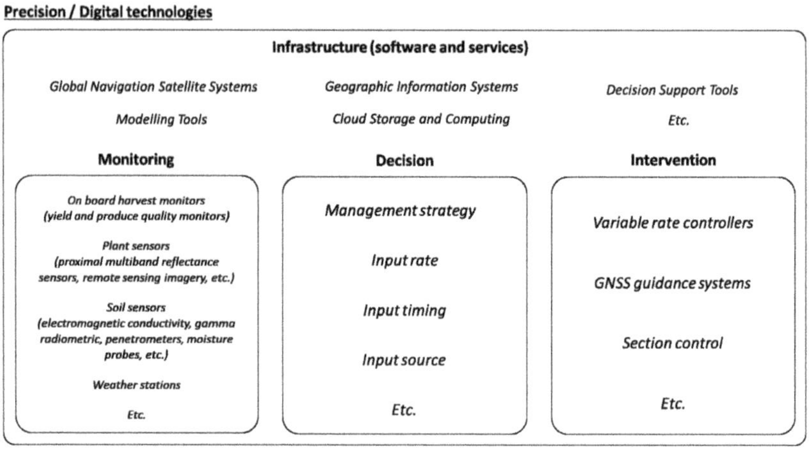

Figure 3 Map of precision and digital technologies in which monitoring and intervention tools are connected by the decision-making step and supported by software and services infrastructure.

- 'Monitoring' (or 'diagnostic') tools for monitoring crop and soil variability such as crop and soil sensors;
- 'Decision' tools for analyzing sensor and other data to identify areas requiring action and suggest what action to take; and
- 'Intervention' tools for implementing management decisions.

These categories are interconnected and overall supported by software and service infrastructure such as GIS, decision support, and modeling tools and other public or private services such as GNSS and cloud computing.

In the case of PA, these technologies have been used to promote more 'precise' management, with a strong focus on more efficient and targeted input application, both to achieve target yields with less input use (improving profitability) and in part to reduce damaging and wasteful input losses to the environment. This general goal has been translated into the concept of the 4 'R's:

- The right source (applied at);
- The right rate;
- The right time; and
- The right place.

There has been debate about whether to add other criteria such as the right **data** and the right **equipment** needed to make PA effective. A recent focus is on the right **purpose**, adding emphasis on the desired output of production systems. For example, an on-farm decision may change depending on the objectives set for the production system. The 'right purpose' has become increasingly important given the recent focus on the environmental and social sustainability of agriculture. It also promotes a re-examination of what 'efficient resource use' means, how it is measured and optimized. There is now pressure to measure agriculture not just in terms of traditional measures, such as yield, but in terms of reducing its environmental impact and contributing to the delivery of ecosystem services which are of broader benefit to society (e.g. soil carbon sequestration to help mitigate climate change).

4 Analyzing material and energy flows in agriculture

Agricultural production can be understood not only as an economic system but also as thermodynamic and physical systems characterized by nutrient and other material as well as energy flows which may enter the system, are cycled within it, and then flow out again. Understanding these material flows within agriculture is critical to better understanding the way resources are used, their impacts and how their use can be improved to optimize system function to meet economic, environmental, or social goals. A good example is understanding energy flows in agriculture, including tracking the sources and levels of environmentally damaging and unsustainable fossil fuel use and ways this use can be reduced.

Digital on-farm data can move from enabling more precise, site-specific management to a broader tracking of inputs and outputs to the system. Data collected from monitoring inputs (e.g. via variable rate controllers) and from monitoring outputs (e.g. harvest data through yield monitors) can provide information to monitor the agronomic system from multiple environmental and other perspectives in terms of mass and energy balances.

In physical terms, biomass production systems such as agriculture are energy transformation units which convert inputs from a primary to a secondary form. In the energy transformation process (e.g. solar energy used in plant photosynthesis, fossil fuel energy incorporated in fertilizer production, and diesel oil consumed by agricultural machines), there will be losses due to the inherent inefficiency of the system (e.g. leaching, drift, and volatilization) and in accordance with the second law of thermodynamics (entropy, i.e. the irreversibility of processes and the unavailability of useful energy).

In analyzing a system, a prerequisite is to measure inputs according to standard unit values. In evaluating agriculture as an economic system, unit values may include $/L, $/kg, $/unit etc., while joules are used for energy evaluation. The measure mass of carbon dioxide equivalent is used to measure and compare the global warming potential of different greenhouse gases. Assigning values to different inputs (e.g. $, MJ, kg CO_2-eq, etc.) makes it possible to assess their relative roles in the overall system and how effectively it operates.

There are unfortunately relatively few studies presenting material flow data in agriculture. The value of some studies is also limited by not clearly defining system boundaries or how the material flows were determined. An example is calculating energy flows based on a country's GDP (gross domestic product) and total energy consumption. While in the case of agriculture this may account for direct energy use (e.g. in the application of a given amount fertilizers and pesticides), it may neglect the indirect associated energy costs (e.g. those involved in input manufacture, fuel and labor needed for application, and machinery depreciation).

The following steps are recommended for determining material flows:

- Clearly mapping the production system to be evaluated and its boundaries;
- Determining the material flow of directly applied inputs (e.g. fertilizers, pesticides, and seeds);
- Determining the material flow of the indirectly applied inputs (e.g. fuel consumption, depreciation of machines, and labor); and
- Calculating indicators of interest (e.g. energy balances, material incorporation, and CO_2-eq emission).

Colaco et al. (2020) used such an approach to monitor the production of oranges in two commercial orchards over five consecutive seasons. They used variable rate fertilizer maps coupled with yield maps to evaluated input/output dynamics from three perspectives:

- First, economically, in which production costs, revenue, and profitability were calculated using site-specific information on fertilizer input and orange yield;
- Second, from a physical perspective, where nutrient inputs through fertilizer and orange yield were converted into nutrient mass exported from the system in order to calculate nutrient balances and identify nutrient deficits or surpluses that could be lost from the system into the environment; and
- Finally, the system was also assessed from an energy perspective in which input and output masses were converted into energy equivalent (the energy required to produce each mass unit of input or output) to allow the assessment of the system's energy flow and the amount of energy embodied in the final product.

All indicators within each category (e.g. economic, physical, and energy) were calculated on a site-specific basis and maps were produced so that decision-makers could assess the variability of the system's performance at different parts of the orchard. This analysis allowed the identification of parts of the orchard which were never profitable (during 5 years of evaluation), primarily due to soil constraints. An example of a site-specific decision driven by economic and environmental factors would be for the grower to remove the uneconomic parts of the orchard to improve overall profitability and convert them into wildlife-friendly habitat, delivering environmental and ecological benefits that could offset environmental impacts made elsewhere on the farm.

Another example of the use of digital data for holistic systems analysis is the work of Spekken et al. (2015). The authors calculated the economic and energy costs associated with nonproductive time spent by field machinery in maneuvering when reaching field borders. They analyzed Brazilian sugarcane production for biofuel, a production system that requires not just economic profitability but also a positive energy balance to establish its overall environmental benefit. By assessing the economic and energy performance of the system site-specifically, the authors were able to identify parts of the field which contained row lengths that were too short to justify the energy cost spent in turning field machinery at the field border. These parts of the field could then be removed from crop production and rewilded both to minimize economic losses and optimize environmental benefits from the system.

According to Barros et al. (2020) there have been a growing number of studies of material, nutrient, and energy flows in agricultural systems, mainly since 2016 and concentrated in European countries. A few illustrative examples from around the world are listed here.

Veiga et al. (2015) assessed the energy return on investment (EROI) for a range of horticultural and perennial crops in Brazil in producing products such as oil and starch for food processing; and Sipert and Cohim (2020) modeled phosphorus flows at a national level in Brazil, concluding that 40% of the phosphorus applied remained in the soil for use by crops, with the remainder lost, including 16% lost through erosion and runoff to watersheds. Tamura and Fujie (2014) assessed loss ratios of livestock manure and potential nutrients from manure that could be recycled to pasture and sugarcane fields. They also proposed the use of other waste materials such as sugarcane tops and straw to produce silage for beef cattle as a basis for more circular systems in which waste outputs could be repurposed as inputs going back into the system. Hu et al. (2015) performed a similar analysis in South Central China, tracking both inputs such as pesticides and fertilizers as well as outputs/emissions such as straw burning, livestock manure, pesticides and fertilizer leaching. In Uganda, Lederer et al. (2015) identified high soil deficits for nitrogen, phosphorus, and potassium, proposing use of sewage and solid waste reduce this deficit and enhance yields. Other studies have looked at global system energy and nutrient flows. Oita et al. (2015) used a multi-regional input-output (MRIO) assessment to track global nitrogen flows, identifying the main trading partners and commodities responsible for embodied nitrogen. Nesme et al. (2018) have estimated the net budget of phosphorus around the world based on FAOSTAT data. These approaches should help to provide information about traceability which is trustworthy, comprehensive, and detailed enough to be able to guide policy response and to support stakeholders and policy makers in monitoring and governing global trade-flows and their undesired impacts (Bruckner et al., 2019).

5 Digital agriculture and the sustainability agenda

Agriculture has always been closely linked with environmental issues. One of the biggest current challenges facing the world is to feed a still growing population without causing potentially catastrophic damage to the environment (Conway, 2019; Vasylieva, 2019). The term 'sustainable agriculture' can be traced back to the Brundtland Report published in 1987 and is based on the paradigm of sustainable development (Pretty, 1995). Sustainable agriculture seeks the efficient management of resources in order to maintain production to meet the need for food and other materials but without causing negative environmental and other impacts. As the FAO indicates, sustainable agriculture must meet the needs of both current and future generations while ensuring profitability, environmental health, and social and economic equality (https://www.fao.org/sustainability/en/). This requires redesigning agricultural systems to reduce negative impacts while meeting these wide-ranging environmental and social objectives (Khatounian, 2001; Uphoff, 2002; Altieri, 2009). There have been many studies of how to achieve a 'triple bottom line' of beneficial economic, environmental, and social outcomes in agriculture (Bird and Ikerd, 1993; Paterniani, 2001; Kitamura, 2003; Bordonal et al., 2018; Veiga et al., 2018; Medeiros et al., 2020).

In 2015, a 2030 Agenda for Sustainable Development was adopted by all 193 Member States of the United Nations. The Agenda included 17 Sustainable Development Goals (SDGs) backed up by 169 targets and over 230 indicators to ensure progress in achieving the SDGs (UN, 2015; Bennich et al., 2020). Particular SDGs stand out in relation to agriculture. SDG 2 is centered on the objective of 'zero hunger and sustainable agriculture'. SDG 12 concerns 'sustainable consumption and production'. More indirectly SDG 9 (industry, innovation and infrastructure) supports developments in technology which enable more sustainable agricultural production.

Digital technologies can contribute to achieving the SDGs at three levels:

- At the farm level, by improving the efficiency with which resources are used, increasing productivity and reducing negative environmental impacts;

- At the value chain level by improving efficiency in the supply chain and by enabling the traceability needed to measure and certify good practice; and
- At the level of the public sphere by giving governments and others the information to develop policy, create regulatory frameworks, and monitor compliance.

At the field level, DA can help to monitor GHG emissions and other negative environmental impacts of farming operations, reduce, and optimize input use (reducing impacts caused by inputs as well as emissions related to their manufacture and application) while still maintaining output and potentially improving farm productivity (Waldrop et al., 2004; Angers and Eriksen-Hamel, 2008; Balafoutis et al., 2017).

At the value chain level, digital technologies make it easier to measure the impact of particular processes/links in the supply chain and provide traceability back to a particular process or actor. This makes it easier to enforce regulatory compliance and support certification schemes promoting good practice. An example is linking on-farm practices promoting more efficient resource use, soil health and biodiversity conservation practices to measurable reduction in CO_2 emissions or soil organic carbon sequestration. These can then enable access to government schemes promoting low carbon farming for its broader environmental benefits or to commercial schemes allowing trading of carbon credits.

At the level of the public sphere, digital technologies can help governments to identify the impacts of existing farming practices and design more environmentally, socially, and economically friendly alternatives to inform policy and regulation. An international example is the Amazon Soy Moratorium, a trade agreement between global producers and investment institutions to restrict soybean cultivation and export solely to existing agricultural land in Brazil (with a ban on soy produced as a result of post-2008 deforestation). To enforce this agreement, satellite imagery is used to monitor the extent of land used for soybean cultivation along with deforestation monitoring data produced by INPE-PRODES.

6 Conclusion and future trends

Since the beginning of agriculture, human development has been marked by cycles of increased production and efficiency, leading to demographic expansion and technological development which, in turn, feed back into further improvements in production and efficiency. Digital technologies form the basis of the latest (and current) revolution in agriculture in the endless search for increased production, efficiency, and sustainability.

With their beginnings in precision agriculture, digital agricultural technologies focused initially on ways of optimizing resource use in farming operations to maintain or improve output at lower cost and higher profitability. Since then, these technologies have needed to expand both to cover greater scales (e.g. monitoring landscapes or regions) and to encompass environmental and social as well as economic targets. Digital technologies are now being deployed beyond the economic realm to monitor farm energy and carbon footprints.

Future research is needed to further develop digital agricultural technologies in two critical areas:

- Holistic system monitoring in which physical farm inputs and outputs are mapped (e.g. using sensors mounted on field machinery) and system performance is assessed using a range of economic, environmental and other indicators; and
- The use and integration of digital technologies at multiple levels, including at the farm level for increased resource use efficiency, along the value-chain to support traceability, environmental certification, and incentive schemes and for regional landscape monitoring to frame public policies.

These approaches are all critical in the successful application of digital agriculture to help achieve global sustainably goals.

7 References

Altieri, M. (2009). *Agroecologia: A Dinâmica Produtiva da Agricultura Sustentável*. Editora da Universidade/UFRGS, Porto Alegre, BR.

Angers, D. A. and Eriksen-Hamel, N. S. (2008). Full-inversion tillage and organic carbon distribution in soil profiles: a meta-analysis. *Soil Science Society of America Journal* 72(5), 1370–1374. https://doi.org/10.2136/sssaj2007.0342.

Anonymous (2022). Precision agriculture definitions. Available at: http://www.grap.udl.cat/en/presentation/pa_definitions.html.

Balafoutis, A., Beck, B., Fountas, S., Vangeyte, J., Van der Wal, T., Soto, I., Gómez-Barbero, M., Barnes, A. and Eory, V. (2017). Precision agriculture technologies positively contributing to GHG emissions mitigation, farm productivity and economics. *Sustainability* 9(8), https://doi.org/10.3390/su9081339.

Barros, M. V., Salvador, R., de Francisco, A. C. and Piekarski, C. M. (2020). Mapping of research lines on circular economy practices in agriculture: from waste to energy. *Renewable and Sustainable Energy Reviews*, 131, 109958. https://doi.org/10.1016/j.rser.2020.109958.

Bennich, T., Weitz, N. and Carsen, H. (2020). Deciphering the scientific literature on SDG interactions: a review and reading guide. *Science of the Total Environment* 728, 138405. https://doi.org/10.1016/j.scitotenv.2020.138405.

Bentley, J. H. (2013). Environmental crises in world history. *Procedia – Social and Behavioral Sciences* 77, 108–115. https://doi.org/10.1016/j.sbspro.2013.03.067.

Bird, G. W. and Ikerd, J. (1993). Sustainable agriculture: a twenty-first-century system. *The Annals of the American Academy of Political and Social Science* 529(1), 92–102. https://doi.org/10.1177/0002716293529001009.

Bois, G. (1985). Population, resources et progrès technique dans un village du Mâconnais (Xe – XVIIIe siècles). In: Gourou, P. and Etienne, G. (Eds). *Des Labours de Cluny à la Revolution Verte. Techniques Agricoles et Population*. PUF, Paris, pp. 25–39.

Bongiovanni, R. and Lowenberg-deBoer, J. (2004). Precision agriculture and Sustainability. *Precision Agriculture* 5(4), 359–387.

Bordonal, R. O., Carvalho, J. L. N. C., Lal, R., Figuiredo, E. B., Oliveira, B. G. O. and Scala Jr, N. L. (2018). Sustainability of sugarcane production in Brazil. A review. *Agronomy for Sustainable Development* 38(2), https://doi.org/10.1007/s13593-018-0490-x.

Bruckner, M., Häyhä, T., Giljum,S., Maus, V., Fischer, G., Tramberend, S. and Börner, J. (2019). Quantifying the global cropland footprint of the European Union's non-food bioeconomy. *Environmental Research Letters* 14, 045011. https://doi.org/10.1088/1748-9326/ab07f5.

Cassman, K. G. (1999). Ecological intensification of cereal production systems: yield potential, soil quality, and precision agriculture. *Proceedings of the National Academy of Sciences of the United States of America* 96(11), 5952–5959.

Chlingaryan, A., Sukkarieh, S. and Whelan, B. (2018). Machine learning approaches for crop yield prediction and nitrogen status estimation in precision agriculture: a review. *Computers and Electronics in Agriculture* 151(1), 61–69.

Colaço, A. F. and Bramley, R. G. V. (2018). Do crop sensors promote improved nitrogen management in grain crops? *Field Crops Research* 218, 126–140. https://doi.org/10.1016/j.fcr.2018.01.007.

Colaco, A. F., Pagliuca, L. G., Romanelli, T. L. and Molin, J. P. (2020). Economic viability, energy and nutrient balances of site-specific fertilization for citrus. *Biosystems Engineering* 200(1), 138–156.

Colaço, A. F., Richetti, J., Bramley, R. G. V. and Lawes, R. A. (2021). How will the next-generation of sensor-based decision systems look in the context of intelligent agriculture? A case-study. *Field Crops Research* 270(1), 108205. https://doi.org/10.1016/j.fcr.2021.108205.

Conway, G. (2019). *The Doubly Green Revolution – Food for All in the 21st. Century*. Cornell University Press, New York.

Cook, S., Jackson, E. L., Fisher, M. J., Baker, D., and Diepeveen, D. (2021). Embedding digital agriculture into sustainable Australian food systems: pathways and pitfalls to value creation. *International Journal of Agricultural Sustainability* 20(3), 346–367.

Dong, D., Wang, C., Van Zwieten, L., Wang, H., Jiang, P., Zhou, M. and Wu, W. (2020). An effective biochar-based slow-release fertilizer for reducing nitrogen loss in paddy fields. *Journal of Soils and Sediments* 20, 3027–3040. https://doi.org/10.1007/s11368-019-02401-8.

FAO (2017). Faostat data. Available at: http://www.fao.org/faostat/en/#data. (Accessed on 08 May 2018).

Franco, H. C. J., Trivelin, P. C. O., Faroni, C. E., Vitti, A. C. and Otto, R. (2008). Aproveitamento pela cana-de-açúcar da adubação de plantio. *Revista Brasileira de Ciência do Solo* 32(September), 2763–2770.

Global Yield Gap and Water Productivity Atlas (2022). Available at: http://www.yieldgap.org. (Accessed on 23 July 2022).

Hamilton, M. J., Lobo, J., Rupley, E., Youn, H. and West, G. B. (2016). The ecological and evolutionary energetics of hunter-gatherer residential mobility. *Evolutionary Anthropology* 25(3), 124–132. https://doi.org/10.1002/evan.21485.

Hémery, D., Debier, J. C. and Deléage, J. P. (1993). *Uma história da energia. (Les servitude de la puissance: une histoire de l'energie. Trad.: Sérgio de Salvo Brito)*. Editora da Universidade de Brasília, Brasília, 447 p.

Hu, R., Huang, X., Huang, J., Li, Y., Zhang, C., Yin, Y., Chen, Z., Jin, Y., Cai, J. and Cui, F. (2015). Long-and short-term health effects of pesticide exposure: A cohort study from China. *PLoS ONE* 10 (6), e0128766. https://doi.org/10.1371/journal.pone.0128766.

Khatounian, C. A. (2001). *A Reconstrução Ecológica da Agricultura*. Agroecológica, Botucatu, BR.

Kitamura, P. C. (2003). Agricultura sustentável no Brasil: avanços e perspectivas. *Ciência e Ambiente* 27(1), 7–28.

Klompenburg, T. V., Kassahun, A. and Catal, C. (2020). Crop yield prediction using machine learning: a systematic literature review. *Computers and Electronics in Agriculture* 177, 105709. https://doi.org/10.1016/j.compag.2020.105709.

Lambert, D. and Lowenberg-deBoer, J. (2000). *Precision Agriculture Profitability Review*. Site-specific Management Center School of Agriculture Purdue University. Available at: https://ag.purdue.edu/SSMC/Frames/newsoilsX.pdf.

Liakos, K. G., Busato, P., Moshou, D., Pearson, S. and Bochtis, D. (2018). Machine learning in agriculture: a review. *Sensors* 18(8), 1–29. https://doi.org/10.3390/s18082674.

Lipinski, B., Clowes, A., Goodwin, L., Hanson, C., Swannell, R. and Mitchell, P. (2017). SDG target 12.3 on food loss and waste:2017 progress report.Washington, DC: Champions 12.3.https://champions123.org/sites/default/files/2020-09/champions-12-3-2017-progress-report.pdf.

Marchetti, C. (2009). On energy and agriculture: from hunting gathering to landless farming. *Global Bioethics* 22(1–4), 29–39. https://doi.org/10.1080/11287462.2009.10800682.

Medeiros, C. A. B., Bueno, Y. M., Sa, T. D. A., Vidal, M. C., Espindola, J. and A. A. (2020). *Zero Hunger: Contributions of Embrapa*. Embrapa, Brasília, SP.

Minussi Winck, J. E., Tagliapietra, E. L., Schneider, R. A., Inklman, V. B., Nara, M. D., Savegnago, C., Paula, L. S., Da Silva, M. R., Zanon, A. Jr. and Streck, N. A. (2023). Decomposition of yield gap of soybean in environment × genetics × management in Southern Brazil. *European Journal of Agronomy* 145, 126795.

Misara, R., Verma, D., Mishra, N., Rai, S. K. and Mishra, S. (2022). Twenty-two years of precision agriculture: a bibliometric review. *Precision Agriculture* 23(6), 2135–2158. https://doi.org/10.1007/s11119-022-09969-1.

Naylor, R. L. (1996). Energy and resource constraints on intensive agricultural production. *Annual Review of Energy and the Environment* 21(1), 99–123. https://doi.org/10.1146/annurev.energy.21.1.99.

Nesme, T., Metson, G. S. and Bennett, E. M. (2018). Global phosphorus flows through agricultural trade. *Global Environmental Change* 50, 133-141.

Passet, R. (1984). *La Pensée Marxienne Sous les Feux de l Athermodynamique. Mélanges J. Weiller*. Economica, Paris.

Paterniani, E. (2001). Agricultura sustentável nos trópicos. *Estudos Avançados* 15(43), 303–326. https://doi.org/10.1590/S0103-40142001000300023.

Pretty, J. N. (1995). Participatory learning for sustainable agriculture. *World Development* 23(8), 1247–1263. https://doi.org/10.1016/0305-750X(95)00046-F.

Rashmi, I., Roy, T., Kartika, K. S., Pal, R., Coumar, V., Kala, S. and Shinogi, K. C. (2020). Organic and inorganic fertilizer contaminants in agriculture: impact on soil and water resources. *Contaminants in Agriculture*. https://doi.org/10.1007/978-3-030-41552-5_1.

Romanelli, T. L. (2021). Avaliação da sustentabilidade energética de sistemas agrícolas: abordagens teóricas e considerações práticas (Assessment of energy sustainability of agricultural systems: theoretical approaches and practical considerations). In: Philippi Jr, A. and Andreoli, C. V. (Eds.), *Sustentabilidade no Agronegócio ('Sustainability on Agribusiness')*. Editora Manole, Curitiba, PR, Brazil, pp. 155–188.

Rong, L.-B., Gong, K.-Y., Duan, F.-Y., Li, S.-K., Zhao, M., He, J.-Q., Zhou, W.-B. and Qiang, Y. (2021). Yield gap and resource utilization efficiency of three major food crops in the world – A review. *Journal of Integrative Agriculture* 20(2), 349–362.

Rose, D. C. and Chilvers, J. (2018). Agriculture 4.0: broadening responsible innovation in an era of smart farming. *Frontiers in Sustainable Food Systems* 2(1), 1–7.

Schueller, J. K. and Bae, Y. H. (1987). Spatially attributed automatic combine data acquisition. *Computers and Electronics in Agriculture* 2(2), 119–127.

Shepherd, M., Turner, J. A., Small, B. and Wheeler, D. (2020). Priorities for science to overcome hurdles thwarting the full promise of the 'digital agriculture' revolution. *Journal of the Science of Food and Agriculture* 100(14), 5083–5092.

Sipert, S. and Cohim, E. B. (2020). A phosphorus flow analysis of Brazil. *Environmental Engineering Science* 37 (2). https://doi.org/10.1089/ees.2019.0264.

Spekken, M., Molin, J. P. and Romanelli, T. L. (2015). Cost of boundary manoeuvres in sugarcane production. *Biosystems Engineering* 129(1), 112–126.

Stafford, J. (Ed.). (2019). *Precision Agriculture for Sustainability*. Burleigh Dodds Science Publishing, Cambridge.

Tamura, S. and Fujie, K. (2014). Material cycle of agriculture on Miyakojima Island: material flow analysis for sugar cane, pasturage and beef cattle. *Sustainability* 6 (2), 812–835. https://doi.org/10.3390/su6020812.

Tilman, D., Cassman, K. G., Matson, P. A., Naylor, R. and Polasky, S. (2002). Agricultural sustainability and intensive production practices. *Nature* 418(6898), 671–677. https://doi.org/10.1038/nature01014.

Tudge, C. (1998). *Neanderthals, Bandits and Farmers: How Agriculture Really Begins*. Weidenfeld & Nicolson, London. 53p. ISBN 0297842587.

United Nations (2015). Transforming Our World: The 2030 Agenda for Sustainable Development. Department of Economic and Social Affairs. Available at: un.org. (Accessed on 27 August 2022).

Uphoff, N. (2002). The agricultural development challenges we face. In: Uphoff, N. (Ed.), *Agroecological Innovations*. Earthscan Publications Ltd, London, pp. 3-20.

Vasylieva, N. (2019). Improvement of agricultural management: functional comparative approach. *Montenegrin Journal of Economics*. Economic Laboratory for Transition Research (ELIT) 15(1), 227–238. Available at: https://ideas.repec.org/a/mje/mjejnl/v15y2019i1227-238.html.

Veiga, J. P. S., Malik, A., Lenzen, M., Ferreira Filho, J. B. S. and Romanelli, T. L. (2018). Triple-bottom-line assessment of São Paulo state's sugarcane production based on a Brazilian multi-regional input-output matrix. *Renewable and Sustainable Energy Reviews* 82(1), 666–680. https://doi.org/10.1016/j.rser.2017.09.075.

Veiga, J. P. S., Romanelli, T. L., Gimenez, L. M., Busato, P. and Milan, M. (2015). Energy embodiment in Brazilian agriculture: an overview of 23 crops. *Agricultural Engineering, Sci. agric.(Piracicaba, Braz.)* 72 (6). https://doi.org/10.1590/0103-9016-2015-0188.

Waggoner, P. E. (1995). How much land can ten billion people spare for nature? Does technology make a difference? *Technology in Society* 17(1), 17–34. https://doi.org/10.1016/0160-791X(94)00024-8.

Waldrop, M. P., Zak, D. R., Sinsabaugh, R. L., Gallo, M. and Lauber, C. (2004). Nitrogen deposition modifies soil carbon storage through changes in microbial enzymatic activity. *Ecological Applications* 14(4), 1172–1177.

World Bank (n.d.). Individuals using the Internet (% of population). Available at: https://data.worldbank.org/indicator/IT.NET.USER.ZS. (Accessed on 24 May 2022).

Chapter 3

Key issues in incorporating proximal and remote sensor data into farm decision-making

Adélia M. O. Sousa, Universidade de Évora, MED, CHANGE, EarsLab, Portugal; José R. Marques da Silva, Universidade de Évora, MED, CHANGE, Agroinsider Lda, Portugal; João Serrano, Shakib Shahidian and Duarte Lobo da Silveira, Universidade de Évora, MED, CHANGE, Portugal; Manuela Simões, Universidade Nova de Lisboa, Portugal; Ana Cristina Gonçalves, Maria João P. Caldinhas and Vasco Fitas da Cruz, Universidade de Évora, MED, CHANGE, Portugal; Arilson J. de Oliveira Júnior and Silvia R. Lucas de Souza, São Paulo State University, Brazil; Diogo R. Coelho, Universidade de Évora, MED, Portugal; Patrícia Lourenço, Agroinsider Lda, Portugal; and Fátima F. Baptista, Universidade de Évora, MED, CHANGE, Portugal

1 Introduction

The last few decades have seen a significant evolution in remote and proximal sensor technologies, data digitalization, analysis and communication tools, etc. that have impacted sectors such as agriculture and forestry. Developments include global navigation satellite system (GNSS) for machine location as well as for mapping and soil sampling. Other important developments include commercial remote sensing technologies to study and monitor agricultural crops, variable rate technology (VRT), the development of robots (e.g. drones) and the use of algorithms to speed up decision-making processes. Remote

http://dx.doi.org/10.19103/AS.2023.0132.04

sensing technologies, including satellites, drones and proximal sensors, provide high-resolution imagery and data that enable farmers to monitor the fields with precision. These data allow precision resource management of water, soil, fertilizers and pesticides.

Society is facing several important challenges such as feeding a growing population, the impacts of climate change, water scarcity, poor soil health, preserving biodiversity, etc. Regions such as the EU have transformative policies to decarbonize the economy and become climate-neutral by 2050. Since the agricultural sector contributes up to 30–40% of the global carbon footprint, smart farming can help contribute to reduce greenhouse gas (GHG) emissions and production costs by increasing input use efficiency.

However, effectively implementing smart technologies requires a certain level of technical expertise and skills. The first challenge faced by farmers in adopting these technologies is to overcome the reluctance to change. Overcoming these barriers often involves training support programmes to help farmers to understand the benefits and the applications of smart technologies. Farmers and agricultural professionals need to be trained to use and interpret the high volume of data obtained with these technologies. Interpreting remote sensing data, such as satellite imagery or drone-based data, can be complex. It involves understanding several indices, spectral bands and how they relate to specific agriculture or forestry parameters. Another barrier is the investment and maintenance costs for equipment, including sensors, drones and software. Addressing these challenges often involves a collaborative work between stakeholders, including government agencies, technology providers, agricultural extension services and educational institutions. Training programmes and support services can help farmers and technicians to overcome barriers and harness the full potential of smart technologies and remote sensing in agriculture.

This chapter presents five study cases on the application of smart technologies in farming and forestry systems. The first (Section 2) of these study cases presents the results of a long-term study to calibrate a Grassmaster II capacitance probe to estimate pasture productivity in the Mediterranean *Montado* ecosystem. The second (Section 3) presents a methodology for identifying yield zones within an olive grove based on Sentinel-2 satellite data. The third case study (Section 4) provides an overview of different remote sensing sensors, data and methodologies used for estimating forest biomass. The fourth case study (Section 5) assesses the carbon footprint of horticultural crops such as potato, onion, carrot, melon and watermelon. The final (Section 6) focuses on precision livestock farming, specifically a mobile application focused on thermal comfort.

2 Proximal sensor calibration and validation: evaluating the Grassmaster II probe to estimate pasture productivity

Estimation of pasture productivity at various stages of the growth cycle is an important element in planning and decision-making by the farmer. Pasture productivity affects management of animal grazing in terms of animal stocking and rotation across different fields, calculation of forage needs and feed supplementation required. Measurement of pasture productivity can be carried out by direct or indirect methods. The conventional (direct) method is based on harvesting samples in specific sampling areas, a lengthy and expensive process given the large number of samples necessary to accurately represent a pasture (Sanderson et al., 2001; Cauduro et al., 2006). Since this method is not practical at the farm level, indirect techniques have emerged that provide an estimate of productivity and its spatial variability over large areas in a timely and cost-effective manner (Serrano et al., 2011; Pullanagari et al., 2013).

A number of optical sensors have been developed such as 'Crop Circle', 'Yara N-Sensor', 'GreenSeeker' or 'OptRx' (Serrano et al., 2019). These sensors typically use reflectance in visible and near-infrared wavelengths to infer plant properties. Vegetative indices use two or more spectral regions, combined in an equation, to infer plant properties. The most widely used index is the normalized difference vegetative index (NDVI) which tends to peak when plant productivity is highest (Cicore et al., 2016; Schaefer and Lamb, 2016).

Using an alternative approach, a New Zealand company, Novel Ways Electronic, has developed a capacitance probe (Grassmaster II), described by Serrano et al. (2011). This probe is based on the pioneering theoretical work of Vickery and Nicol (1982). Capacitance instruments generate an electrical signal of a certain frequency to produce an electric field. The Grassmaster II probe senses this electrical field in an area approximately 100 mm in diameter and 400 mm in length (Sanderson et al., 2001). The electrical field generated by the probe is modified by the pasture close to it, allowing the probe to measure the capacitance (ability to store electrical charge) of the air-herbage mixture. The capacitance value is proportional to the water content of the grass. Because dry matter content is highly correlated with water content, changes in capacitance can be related to pasture yield. Linear calibration equations are used to correlate readings with measured dry matter production (Sanderson et al., 2001; Serrano et al., 2011). The equation used in Grassmaster II is:

$$DM = 0.72 \times CMR - 2200$$

where DM is the estimated productivity of the pasture (in kilogram of dry matter per hectare), and CMR is the capacitance measured by the probe.

The relationship between pasture productivity and capacitance measurement is influenced by factors such as the mixture of plant species and their phenological stage (Virkajarvi, 1999; Sanderson et al., 2001; Zanine et al., 2006). According to Hanna et al. (1999), seasonal adjustments to calibration equations are necessary as the moisture content of pasture vegetation varies with season and live/dead material ratio. This variability requires calibration tests to take account of these variables. In addition, the growth of pastures in a particular area is affected by factors such as soil fertility, grazing strategy and especially rainfall pattern, which introduces inter-annual variation in terms of both growth and development, e.g. in an area such as the dryland pastures of the Alentejo area of Portugal which is part the *Montado* ecosystem. The *Montado* ecosystem is a unique savannah-like woodland of cork/holm oak trees found in the Mediterranean (Pullanagari et al., 2013). This unique agro-forestry system consists of annual or permanent bio-diverse pastures (composed of different legumes, grasses, composites and other species) grazed by sheep or cattle in rotational or permanent systems. This section discusses the results of tests carried out in a long-term study carried out between 2007 and 2019 to calibrate and validate a capacitance probe (Grassmaster II) to estimate pasture productivity in this system.

2.1 Materials and methods

The experiments were carried out in two phases:

- Calibration phase - between 2007 and 2018, included measurements at three fields located in the Évora District; and
- Validation phase - between February and March 2019, included measurements at four parcels located in the Alentejo region.

Pasture sampling in the calibration phase was carried out at each location, with repetitions through the growth cycle (between February and June), totalling 1411 samples, spread across three fields:

- 'Mitra' (726 samples, taken during 6 years: 2007, 2013 and between 2015 and 2018);
- 'Revilheira' (199 samples, taken during 2 years, 2007 and 2013); and
- 'Silveira' (486 samples, taken during 4 years, between 2013 and 2016).

Pasture sampling in the validation phase was carried out between February and March 2019 in four fields representative of pastures in this region ('Cubillos', 'Grous', 'Murteiras' and 'Tapada').

The sampling process consisted of measuring pasture capacitance with the Grassmaster II probe (Fig. 1), followed by placing a metal rim with an area of 0.1 m^2 (0.25 m × 0.40 m in size; Fig. 1b) over the pasture. The pasture contained in

Figure 1 Pasture sampling process: (a) measuring pasture capacitance with the Grassmaster II probe; (b) pasture sampling area; (c) laboratory procedures.

this area was cut with electric shears and preserved in numbered plastic bags. Once in the laboratory, the pasture sample was weighed (Fig. 1c) to establish total biomass produced by unit area (GM in kg ha^{-1}), dried in an oven (72 h at 65°C) and weighed again to establish productivity in terms of dry matter per unit of area (DM in kg ha^{-1}), according to standard methodology.

The statistical analysis of the results included a descriptive analysis with calculation of average and standard deviation (SD) of each data set. Analysis of the linear correlations between capacitance (CMR) measured by the probe and values of pasture productivity (pasture green matter and pasture dry matter, respectively, GM and DM, in kg ha^{-1}) was carried out with MSTAT-C software, with a significance level of 95% ($P < 0.05$). The data were organized and analysed by pasture moisture content (PMC) classes. The Pearson (*r*) correlation coefficient was used to measure the degree of correlation or the linear dependence between variables and the coefficient of determination (R^2) to measure the proportion of the total variation of the dependent variable explained by the variation of the independent variable. The rigor of the resulting regression models was evaluated by the absolute value of the root mean square error (RMSE) and its relative value (CV_{RMSE}). This statistical parameter measures the average magnitude of the error resulting from the estimate.

2.2 Results

Tables 1 and 2 present the average values and the SD of the parameters measured in the monitored pastures, in each PMC class considered, in the calibration and validation phases, respectively. The spatial variability, detected

Table 1 Mean ± standard deviation of the parameters measured in the pastures of the three fields used in the calibration phase

PMC class (%)	*n*	CMR	GM (kg ha^{-1})	DM (kg ha^{-1})	PMC (%)
>85	236	8772 ± 2887	23696 ± 17570	2923 ± 2067	87.6 ± 2.3
80-85	331	7523 ± 2370	14351 ± 9374	2444 ± 1564	82.8 ± 1.4
75-80	231	6488 ± 2354	11587 ± 8753	2561 ± 1930	77.8 ± 1.4
70-75	187	6428 ± 2429	9950 ± 6712	2729 ± 1834	72.5 ± 1.4
65-70	143	5727 ± 2624	8645 ± 9825	2775 ± 1756	67.9 ± 6.2
60-65	118	5645 ± 1726	8389 ± 5892	3119 ± 2202	62.8 ± 1.4
<60	165	4586 ± 1143	7023 ± 5089	4586 ± 1143	50.1 ± 11.1
>80	567	8043 ± 2668	17119 ± 11341	2570 ± 1657	84.8 ± 3.0
70-80	418	6461 ± 2385	10854 ± 7939	2636 ± 1887	75.4 ± 3.0
<70	426	5262 ± 1631	7946 ± 5509	3175 ± 2313	59.6 ± 10.5

PMC pasture moisture content; *n* number of samples; CMR 'Grassmaster II' measurements; GM green matter; DM dry matter.

Table 2 Mean ± standard deviation of the parameters measured in the pastures of the four fields used in the validation phase

PMC class (%)	*n*	CMR	GM (kg ha^{-1})	DM (kg ha^{-1})	PMC (%)
>80	48	8882 ± 2815	17629 ± 9608	2898 ± 1394	82.9 ± 1.8

PMC pasture moisture content; *n* number of samples; CMR 'Grassmaster II' measurements; GM green matter; DM dry matter.

Table 3 Linear regression equations proposed to estimate pasture green matter yield (GM) based on 'Grassmaster II' measurements (CMR), for each pasture moisture content (PMC) class considered

PMC classes (%)	*n*	Linear equations	*r*	RMSE (kg ha^{-1})	CV_{RMSE} (%)
>85	236	3.3319*CMR-8225.4	0.759**	5855	24.7
80-85	331	3.0236*CMR-8395.7	0.765**	4597	32.0
75-80	231	2.3271*CMR - 3512.4	0.626*	6813	58.8
70-75	187	1.4044*CMR+922.48	0.508*	5766	57.8
65-70	143	1.4915*CMR+103.8	0.477*	4852	56.1
60-65	118	1.46*CMR+146.37	0.428*	5303	63.2
<60	165	2.1234*CMR-2714.2	0.445*	4460	63.5
>80	567	3.2963*CMR-9392.7	0.775**	4806	28.1
70-80	418	1.9042*CMR-1449.2	0.572*	6839	63.0
<70	426	1.5881*CMR-411.24	0.470*	4837	60.9

PMC, pasture moisture content; *n*, number of samples; CMR, 'Grassmaster II' measurements; GM, green matter; DM, dry matter; *r*, correlation coefficient; RMSE, root mean square error; CV_{RMSE}, coefficient of variation of root mean square error; *correlation significant at the 0.05 level; **correlation significant at the 0.01 level.

by the high coefficients of variation of the pasture productivity, GM and DM (around 50-80%; Tables 1 and 2) and 'Grassmaster II measurements' (CV of around 30%) show variations suitable for precision agriculture (PA) applications.

More than 40% of measurements were performed with PMC >80%, i.e., in the early stages of the pasture vegetative cycle, usually in the months of February and March (Serrano et al., 2019). This is the period when the probe has the highest sensitivity given its operating principle (essentially in responding to wet biomass) (Hanna et al., 1999). According to the manufacturer of the probe, water has the greatest effect on the probe signal. The probe does not provide any readings e.g. when the pasture is dry (during summer) (Serrano et al., 2019).

Tables 3 and 4 show, respectively, the linear regression equations between GM and CMR and between DM and CMR, for each PMC class considered. For each regression equation, the correlation coefficient (*r*) and the root mean error (RMSE) are presented as absolute values and as percentages (CV_{RMSE}). The highest correlation coefficients and the lowest CV_{RMSE} (better predictability)

Table 4 Linear regression equations proposed to estimate pasture dry matter yield (DM) based on 'Grassmaster II' measurements (CMR), for each pasture moisture content (PMC) class considered

PMC classes (%)	*n*	Linear equations	*r*	RMSE (kg ha^{-1})	CV_{RMSE} (%)
>85	236	0.4592*CMR-1281.3	0.750**	844	28.8
80-85	331	0.4983*CMR-1304.5	0.755**	765	31.3
75-80	231	0.5065*CMR-725.1	0.618*	1515	59.2
70-75	187	0.3819*CMR+274.19	0.506*	1577	57.8
65-70	143	0.4829*CMR+9.1169	0.479*	1562	56.3
60-65	118	0.5279*CMR+138.59	0.414*	1996	64.0
<60	165	0.7052*CMR+327.74	0.297ns	2581	72.5
>80	567	0.4656*CMR-1174.8	0.750**	763	29.7
70-80	418	0.4483*CMR-260.57	0.567*	1543	58.5
<70	426	0.4222*CMR+953.26	0.298ns	2122	66.8

PMC pasture moisture content; *n* number of samples; CMR 'Grassmaster II' measurements; GM green matter; DM dry matter; *r* correlation coefficient; RMSE root mean square error; CV_{RMSE} coefficient of variation of root mean square error; *-correlation significant at the 0.05 level; **correlation significant at the 0.01 level; ns not significant.

occur for PMC >80%. As PMC decreases, these indicators deteriorate: the correlation coefficient decreases and the CV_{RMSE} increases. For this reason, the analysis of the data grouped by PMC classes (>80%; 70–80% and <70%) was also considered in this study, which is in agreement with studies by Reese et al. (1980), Karl and Nicholson (1987) and Hirata (2000).

According to Jamieson et al. (1991), such results can be considered acceptable (CV_{RMSE} of 28.1% and 29.7%, respectively, for GM and DM), especially considering that the application of sensor techniques to evaluate the existing variability is difficult on permanent grassland with diverse species, plant spacing, morphology and colour (Schellberg et al., 2008). The situation becomes more complex when grazing animals are involved due to dynamic interactions between plants and animals (Ogura and Hirata, 2001). This degree of uncertainty associated with calibration and validation of this capacitance probe was also noted by other authors (Murphy et al., 1995; Virkajarvi, 1999; Cauduro et al., 2006).

Figure 2 shows the relation between pasture productivity measured in 2019 in four experimental fields ('Cubillos', 'Grous', 'Murteiras' and 'Tapada' -- validation phase) and pasture productivity prediction based on the equations obtained in calibration phase for PMC >80%, proving a very acceptable degree of confidence (CV_{RMSE} of 24.4% and 20.4%, respectively to GM and DM). These results are very promising, especially given the heterogeneity of biodiverse pastures. Each pasture is a different ecosystem, with specific characteristics,

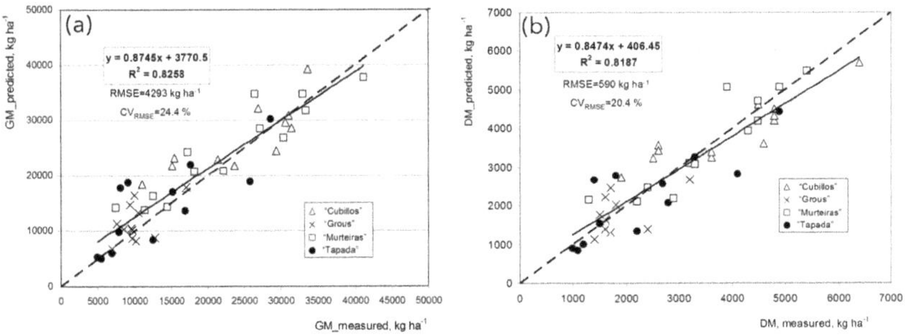

Figure 2 Relation between pasture productivity measured ((a) green matter; (b) dry matter (DM)) in 2019 in four experimental fields ('Cubillos' 'Grous' 'Murteiras' and 'Tapada' - validation phase) and pasture productivity predicted based on the equations obtained in calibration phase for pasture moisture content (PMC) >80%.

which vary according to the different plant species and their vegetative states (Reese et al., 1980; Karl and Nicholson, 1987). Deviations of approximately 20-25% between the measured and the estimated productivity do not, in practice, prevent effective precision management decisions in grazing systems, in particular, for managing the number of animals per hectare. Better results will only be possible in monospecies pastures, e.g. containing only legumes (such as *Trifolium subterraneum*) or grasses (such as *Lolium multiflorum* Lam.), but very unrepresentative of dryland pastures in the Alentejo region (Efe Serrano, 2006).

This long-term study (2007-2019) validates equations for estimating pasture productivity (green and dry matters) based on 'Grassmaster II' capacitance probe measurements. Considering the seasonal changes in PMC, it is clear that the more favorable period for the use of the Grassmaster II probe in dry land biodiverse pastures in Alentejo region of Southern Portugal (PMC > 80%: r = 0.775; $P < 0.01$; RMSE = 4806 kg ha^{-1}; CV_{RMSE} = 28.1% for GM; and r = 0.750; $P < 0.01$; RMSE = 763 kg ha^{-1}; CV_{RMSE} = 29.7% for DM) coincides with the end of winter and beginning of spring (February-March). From late spring onwards (May-June), the rapid drop in PMC adversely affects the measurements and probe accuracy. The estimate of productivity is crucial for the management of the animal grazing, in terms of animal stocking in each field, of animal rotation through different fields, the calculation of the supplementary feed and forage needs for the animals. It can be concluded that the capacitance probe is a useful tool to support decision-making in the management of dryland pasture and grazing.

Funding: This work was funded by National Funds through FCT (Foundation for Science and Technology) under the Project UIDB/05183/2020.

Acknowledgement: Part of this chapter was first published with license by Wageningen Academic Publishers under the title 'Long-term evaluation of the Grassmaster II probe used to estimate productivity of dryland pastures,' in John V. Stafford (ed.), *Precision agriculture '23 (2023)/* 978-90-8686-393-8. It is republished here by license arrangement with Koninklijke Brill NV.

3 Using remote sensing data to identify management zones in olive groves

Hedgerow olive groves (described as such because of their characteristic linear layout) for olive/olive oil production are an important feature of the agricultural landscape of Portugal and many other Mediterranean countries. Optimizing resource use in these systems will have significant potential economic and environmental benefits. Precision olive farming helps to optimize resource use by utilizing information technologies to enable differentiated crop treatments based on yield variability within olive grove plots (Silveira, 2022). Given the lack of available technologies, a study was undertaken to develop a simple and cost-effective methodology capable of identifying yield zones within an olive grove plot.

3.1 Materials and methods

A 22-hectare hedgerow olive grove plot, planted with the Arbequina olive cultivar in 2018 with spacing of 1.35 m × 3.75 m, was selected at Herdade das Romeiras in the Santa Vitória do Ameixial Parish, Estremoz Municipality in Portugal. Satellite images from the Sentinel-2 were collected for the study area at a time when there was no ground cover between the olive tree rows (Fig. 1). This was crucial to prevent natural vegetation from masking the spectral response of the olive grove. Images for bands four and eight, corresponding to red and near-infrared, were downloaded from the Copernicus Open Access Hub for 16 September 2021, on a day with clear sky.

ArcGIS Pro software was used to calculate the NDVI. The principle of the NDVI is based on the fact that vigorous vegetation with active chlorophyll function exhibits high reflectance in the near-infrared band and low reflectance in the visible red band. Radiometric differences between vigorous and senescent vegetation are accentuated by the difference between these bands, allowing for differentiation between types of cover (Fonseca and Fernandes, 2004). The higher the canopy volume of trees in a given area, the higher the NDVI values for pixels representing that same area.

Using ArcGIS Pro, the resulting image pixels were reclassified into five classes organized by the lowest, highest and intermediate NDVI values (Fig. 3). Coordinates for the nine sampling points were recorded and their on-ground locations were identified. Within each of these points, nine olive trees were marked, following the scheme shown in Fig. 4. In total, 81 olive trees were studied (9 locations × 9 olive trees per location).

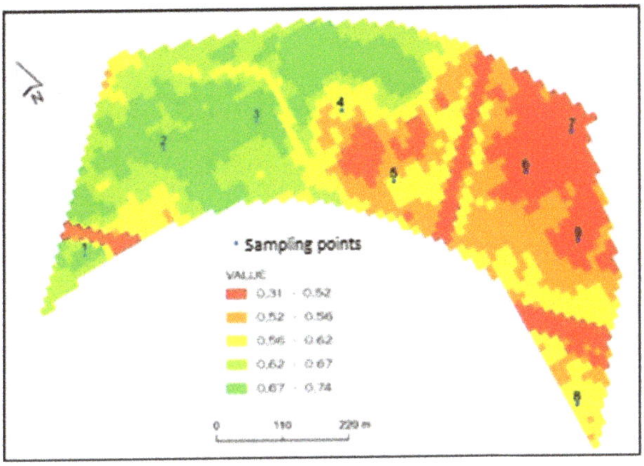

Figure 3 NDVI class map and sampling points location in the Hedgerow olive grove at Herdade das Romeiras, Santa Vitória do Ameixial, Estremoz.

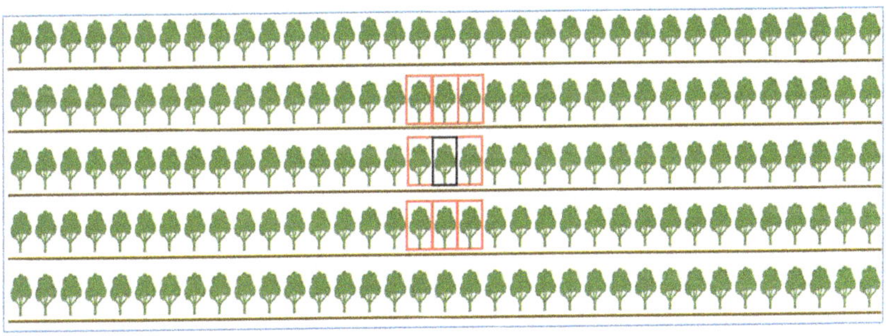

Figure 4 Explanatory diagram of how olive trees were selected at each of the nine sampling points.

Olive harvesting and weighing was carried out in November 2021. Olive trees that constituted moderate outliers within each of the 9 sampling points were excluded from the statistical analysis, resulting in the identification of 12 olive trees (15% of the 81 olive trees under study). Two weeks after harvesting, measurements of the olive tree canopy (height, width and length) were taken using the method suggested by Díez et al. (2016).

3.2 Results

The relationship between the average canopy volume of the olive trees in the study plot in October 2021 and the NDVI on 16 September 2021, can be expressed by the linear function Y = 0.0809 X + 0.2172, with an R^2 of 0.83 (Fig. 5). Since this relationship is statistically significant, by using the function. NDVI imaging from 16 September 2021 and ArcGIS ProR, the spatial distribution of

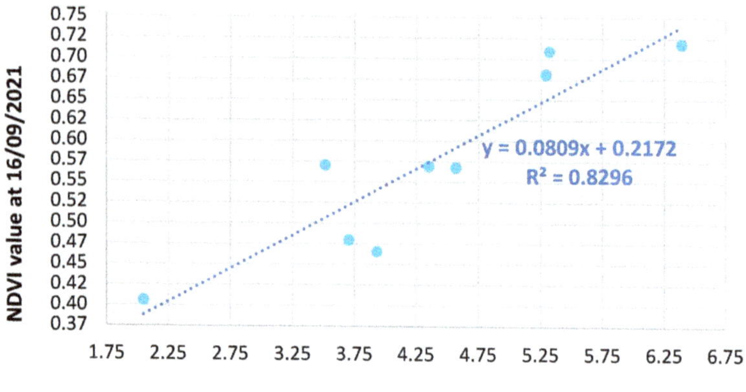

Figure 5 Linear relationship between average olive tree canopy volume and the NDVI value of the representative pixel for those olive trees.

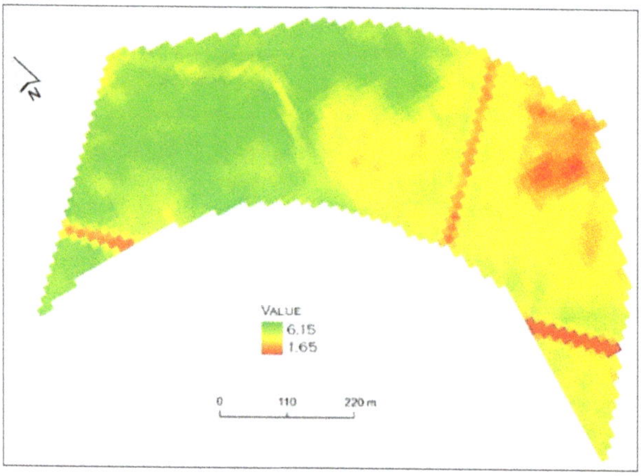

Figure 6 Map of the estimated spatial distribution of the average canopy volume of olive trees (m^3) in the studied olive grove on 16/09/2021.

the average canopy volume of olive trees on the same date could be estimated (Fig. 6).

By cross-referencing the different variables [the canopy volume of the 69 olive trees (81 sampled olive trees - 12 outliers = 69), and the weight of olive production], a statistically significant relationship was observed, with a Pearson correlation coefficient of 0.67. By associating the relationship between NDVI and canopy volume (using a polynomial equation representing the relationship between olive tree canopy volume and olive production weight), a tool for identifying yield areas was created, providing the potential for PA applications with differentiated management of different yield areas.

Based on the data collected, three yield zones were distinguished. A minimum production threshold of 9 kg per tree was defined, based on the average value for the olive grove (9.51 kg $tree^{-1}$). The graphical representation of the function that best fits the collected data (Fig. 7) shows that the defined minimum production corresponds to a minimum canopy volume of 3.56 m^3, while the maximum production (11.20 kg $tree^{-1}$) is associated with a canopy volume of 5.60 m^3.

Using the resulting values from the previous analysis as class interval limits, a map distinguishing the three yield zones was generated through reclassification via ArcGIS ProR, based on the estimated spatial distribution of the average canopy volume of olive trees (Fig. 8). Figure 9 depicts the spatial distribution of the average canopy volume of olive trees, with three volume classes indicated by different colours. The dark green corresponds to the area of olive trees with the highest average canopy volume (>5.60 m^3), covering approximately 1.6 ha on the ground. In this zone, higher resource consumption is predicted Average yield per tree in this zone is not expected to reach maximum values due to the excess canopy. The red-coloured zone represents the area where olive trees have the lowest average canopy volume (<3.56 m^3) and where average production per tree is lower (<9 kg tree−1).

The potential value of these management zones is summarized in Fig. 9, which highlights differences in production requirements between plots. In the dark green area (Fig. 8), which has fertile soil as evident from plant health, it makes sense to reduce the application of fertilizer and water. In the less productive area (red zone; Fig. 8), which has low soil fertility and lower water retention capacity, the relative importance of changes in fertilizer and irrigation management is an issue. The appropriate nutrition strategy should be based on the actual productive potential of the olive grove, not just the theoretical

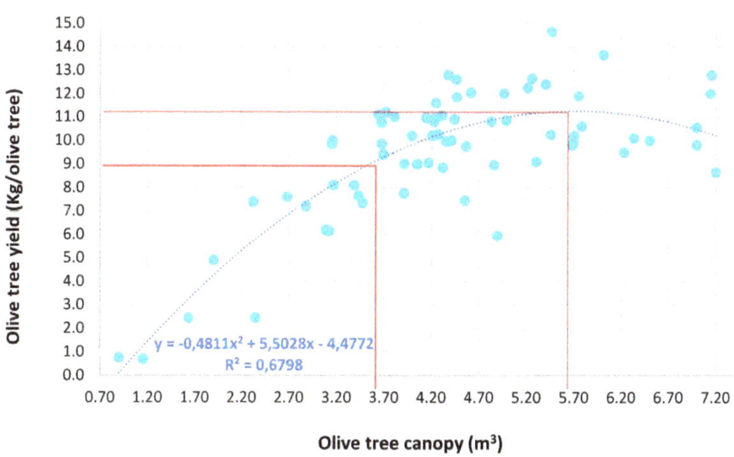

Figure 7 Polynomial relationship between canopy volume and olive yield.

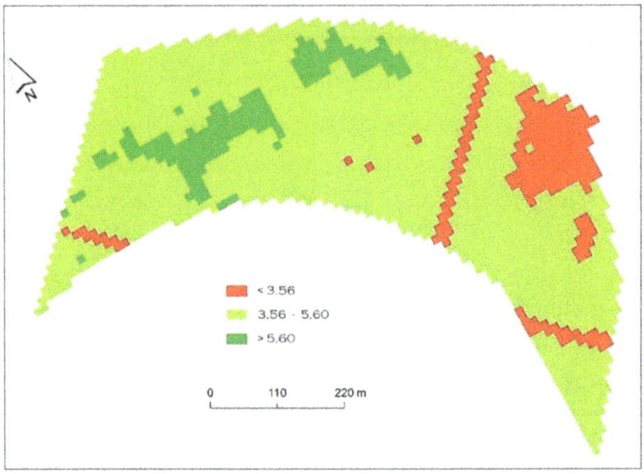

Figure 8 Map of the spatial distribution of average olive tree canopy volume (m^3) considering three volume classes (<3.56 m^3; 3.56 m^3-5.60 m^3; >5.60 m^3). Estimated for 16/09/2021. According to the relationships identified in this study, the volume classes correspond, respectively, to the following yield classes: <9 kg $tree^{-1}$; 9 kg/tree–11.20 kg $tree^{-1}$; <11.20 kg $tree^{-1}$

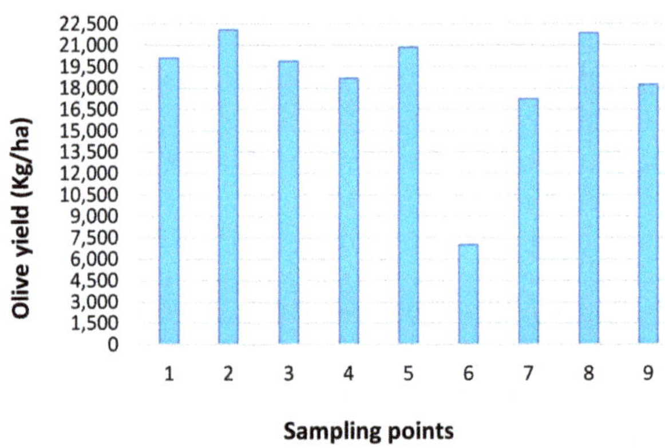

Figure 9 Extrapolation of the average olive production per tree to 1 ha of olive grove.

potential. In most cases, in areas with low soil fertility, such as those highlighted in red, productive potential depends on water availability rather than nutrition. This suggests a reduction in mineral fertilizer application is appropriate, resulting in reduced nutrition in an area corresponding to 3.55 ha (1.60 ha + 1.95 ha), 16% of the total olive grove area.

This study has proposed a simple and cost-effective methodology for distinguishing yield zones within an olive grove plot. NDVI images, constructed from satellite imagery, combined with field data, such as olive production quantity, biometric characteristics and statistical analysis, has enabled the creation of a map that reflects the yield variability of the sample olive grove

(Fig. 8). It was possible to distinguish three yield zones. One zone covers an area of 1.6 ha where the average canopy volume of each olive tree exceeds 5.60 m^3, and the average production per olive tree is below 11.20 kg. Another zone covers an area of 18.45 ha, where the average canopy volume ranges between 3.56 and 5.60 m^3/olive tree, and the average production varies between 9 and 11.20 kg/olive tree. Another area covering 1.95 ha had an average canopy volume of less than 3.56 m^3 with an average yield per tree of less than 9 kg.

Acknowledgements: To the Agricultural Society of Ameixial, with a special thanks to Engineer Henrique Chia, for all the support, kindness and availability. To the professors and colleagues of the Master's in Precision Agriculture Technologies (MTAP). To the Alentejo Olive Oil Sustainability Program (PSAA).

4 Using remote sensing data to estimate aboveground forest biomass

The evaluation and management of forest biomass is crucial for maintaining ecological balance, mitigating climate change, conserving biodiversity and ensuring the well-being of both human and forest systems. Forests are key to the sequestration and storage of carbon (Gonçalves et al., 2021, 2019a). This highlights the need to evaluate, monitor and predict the dynamics of forest biomass in space and time. The quantification of forest biomass can be done using direct or indirect methods. Estimating forest biomass can be undertaken by direct (i.e., destructive) methods with extensive field work, using a sampling design in which a sample of trees is collected, their biomass is weighed and their dry weight determined. With this data it is possible establish mathematical functions between dendrometric variables (e.g. diameter at breast height, height) and biomass, frequently named allometric functions. These functions allow biomass estimates but are specific to forest species. These methods are typically accurate but can be labour-intensive and time-consuming (Watzlawick et al., 2006; Sousa et al., 2015; Zhu and Liu, 2015; Macedo et al., 2018).

With indirect methods, the estimation of forest biomass can be based on other variables related to biomass, such as remote sensing data. These methods are often used for larger areas where direct methods are not feasible. Remote sensing technologies have developed rapidly in recent decades with more varied and higher spatial, radiometric and temporal resolutions (Sousa et al., 2017), enabling periodic monitoring of spatiotemporal changes in forest areas at different scales (local, regional, continental and global). This is crucial for understanding forest dynamics such as growth, mortality and recovery after disturbances. This section reviews different high spatial resolution remote sensing techniques, data and methodologies for estimating forest biomass for four forest species in Alentejo, south of Portugal (*Quercus rotundifolia*, *Quercus suber*, *Pinus pinaster* and *Pinus pinea*), using linear regression and random forest.

Remote sensing data involves the use of both passive (registering the natural energy that is emitted or reflected by objects) and active sensors (emitting their own energy and measuring the energy that is reflected or backscattered) to collect information about the Earth's surface. Remote sensing data from passive or optical sensors used to forest biomass estimation is mainly based on their spectral bands (visible and infrared region of the electromagnetic spectrum). Red and near infrared bands are used to calculate vegetation indices (VI), particularly the NDVI (Rouse et al., 1973). This index is useful for monitoring vegetation growth, assessing crop health and studying changes in vegetation over time.

In addition to satellite spectral bands, Principal Components Analysis (PCA) (Jolliffe, 2002) and the texture index Grey-Level Co-occurrence Matrix (GLCM) (Haralick et al., 1973) have been used to estimate forest biomass. PCA is a mathematical technique to reduce the dimensionality of multispectral or hyperspectral remote sensing data while retaining the most important information, resulting a new set of uncorrelated variables called principal components. Texture analysis refers to the spatial arrangement of pixels or texture elements within an image. It describes the visual patterns or variations in intensity, colour or shape that provide information about the surface properties of the objects in the scene. Both PCA and texture analysis provide additional information beyond traditional spectral indices, which can improve the accuracy of biomass estimation models.

Based on NDVI, some authors (Gonçalves et al., 2017a,b; Macedo et al., 2018a; Sousa et al., 2015) have used a vegetation mask, to identify forest tree crowns by forest species, using multiresolution segmentation and object-oriented image classification (Fig. 10) (Macedo et al., 2018; Lourenço et al., 2021). This mask enables use the spectral response exclusively from the tree crowns, separating the pixels of the tree crowns from the understory vegetation and soil background effects. This, in turn, allows for the elimination of the influence of other land cover types, thereby increasing the accuracy of forest biomass estimation models.

The effectiveness of these techniques depends on factors such as:

- The quality of remote sensing data;
- The choice of features; and
- The robustness of the regression model.

Satellite data spatial resolution is usually divided in three classes: low, medium and high.

- Data with **low spatial resolution (30-1100 m)** is available from National Oceanic Atmospheric Administration-Advanced Very High Resolution

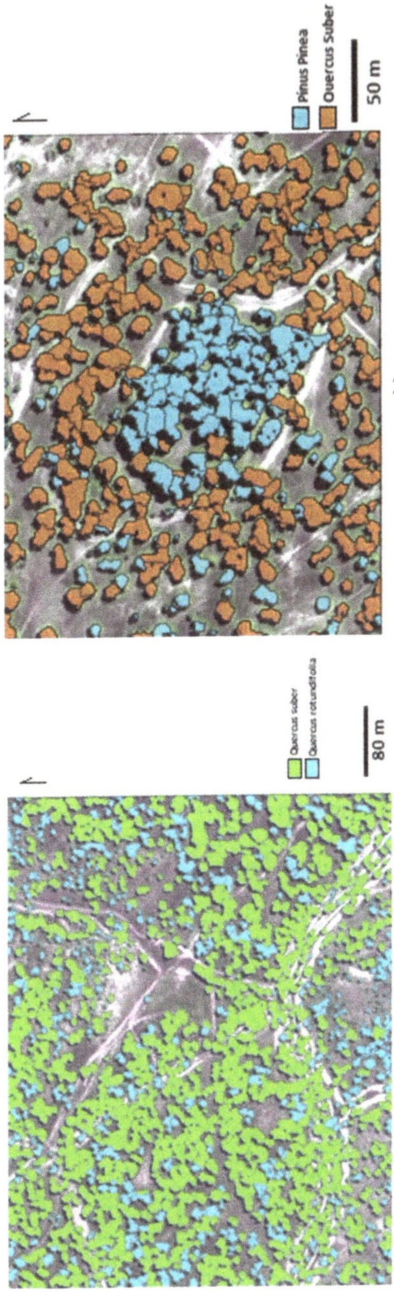

Figure 10 Examples of vegetation mask by forest species, (a) *Quercus suber* and *Quercus rotundifolia*, (b) *Pinus pinea* and *Quercus suber*.

Table 5 Characteristics of satellite/sensors of low, medium and high spatial resolution

Spatial resolution	Satellite/ sensor	Spectral resolution	Spatial resolution (m)	Revisit frequency (days)	Swath width (km)
Low	NOAA-AVHRR	5 bands (0.58–12.50 μm)	1100	1	2922
	MODIS	36 bands (0.40–14.38 μm)	250–1000	1–2	2330
Medium	Landsat 7ETM+/8/9	8 bands–11 bands (0.43–12.51μm)	30 (MS)/15 (PAN)	16	185
	ASTER	14 bands (0.52–11.56 μm)	15/30/90	16	60
	SPOT-4/5/6/7	4 bands (0.50–1.75 μm)	10/20	2–3 days, depending on latitude	60
	Sentinel-2	13 bands (0.43–12.28 μm)	10/20/60	10	290
High	Geoeye-1	4 bands (0.45–0.92 μm)	1.61(MS) - 0.41(PAN)	1–3, depending on latitude	15.2
	IKONOS	4 bands (0.40–4.38 μm)	4 (MS) - 1 (PAN)	3, approximately	11.3
	Quickbird	4 bands (0.43–1.75 μm)	2.4 (MS) - 0.70 (PAN)	1–3.5, depending on latitude	16.5
	WorldView-2/3	8 bands (0.40–1.04 μm)	2.40/2.40 (MS) - 0.50/0.30 (PAN)	1.1–3.7, depending on latitude	16.5
	Pleiades-1	4 bands (0.43–0.93 μm)	1.20 (MS) - 0.30 (PAN)	1	20

Radiometer (NOAA-AVHRR), and Moderate Resolution Imaging Spectroradiometer (MODIS). Data is used for estimating aboveground biomass in large geographic areas (Al-Bakri and Taylor, 2003; Li et al., 2018; Zhang et al., 2018).

- Landsat and Advanced Spaceborne Thermal Emission and Reflection Radiometer (ASTER) satellite data have **medium spatial resolution (10-30 m)**. They are used for regional and country scale biomass estimation (Zheng et al., 2004; Fernández-Manso et al., 2014).
- More recently several satellites with **high spatial resolution (<10 m)** have appeared (Ikonos, Quickbird, WorldView, Geoyeo, Pleiades and Sentinel-2). These data are used to estimate biomass mainly at local or regional scale (Leboeuf et al., 2007; Sousa et al., 2015; Macedo et al., 2018; Wang et al., 2018; Gonçalves et al., 2019; Lourenço et al., 2021).

These are summarized in Table 5.

Conventional multispectral data sets are usually composed of 4-10 spectral bands of relatively large bandwidths (70-400 nm). A more recent development is hyperspectral imaging. Hyperspectral data sets are generally composed of about 100-200 contiguous bands of relatively narrow bandwidths (5-10 nm). Hyperspectral imaging has significantly improved spectral resolution contributing to both identify tree species as well as map large forest areas.

Other new techniques with active sensors include:

- Light Detection and Ranging (LiDAR); and
- Synthetic Aperture Radar (SAR).

These are shown in Table 6. **LiDAR** is based on emission of pulses and measurement of their reflection to generate highly detailed 3D representations of forest structure (Newnham et al., 2015; LaRue et al., 2018). These light sources, combined with other data, generate precise, three-dimensional information about surface characteristics. A LiDAR instrument typically consists of a laser,

Table 6 Characteristics of main active sensor data

Sensor	Data Type	Frequency
LiDAR / ALS	Laser pulse in the optical part of the electromagnetic spectrum (near IR and green) - point cloud	1Hz and 100Hz
SAR / Sentinel-1	Amount energy reflected back after interacting with the Earth surface.	3 MHz - 300 GHz (Band X, C, L, and P) / 5.405 GHz (Band C)

scanner and specialized GNSS receiver. LiDAR data can provide information about the vertical structure of forests which, when compared with other data, can help to assess forest biomass.

The **SAR** system uses microwave signals to penetrate through clouds and vegetation, allowing for year-round data acquisition (Quegan et al., 2000; Townsend, 2002). These systems can transmit and receive electromagnetic waves with different polarizations. The electric field vector of the transmitted wave is in the horizontal (horizontal polarization – H) and in the vertical (vertical polarization – V) plane. There are still radar systems that use both polarizations, referred to as cross-polarization (HV and VH), which involves transmitting with one polarization and receiving orthogonal polarization. The interaction between radar signals and vegetation is influenced by the structure, moisture content and density of the vegetation. Different polarizations provide different perspectives/results, allowing for more accurate biomass estimation (Ghasemi et al., 2011).

Data from passive and active sensors have been combined using more complex regression methods. When it comes to estimating forest biomass using a range of data, several regression methods can be used, from linear regression to machine learning methods. These methods make it possible to create models that relate ground-based biomass estimations and information from remote sensing data.

For linear regression, the dependent variable is the forest biomass and independent variables can be other forest parameters derived from remote sensing data (Gonçalves et al., 2017a,b; Macedo et al., 2018a; Sousa et al., 2015). Machine learning methods include random forests, support vector machines and neural networks. These methods can learn complex relationships between spectral signatures and biomass, enabling accurate biomass prediction for larger areas. It is important to note that the performance of regression methods varies depending on the dataset, satellite data characteristics and the quality and quantity of ground truth biomass estimations. It is recommended to test different methods, compare their results and evaluate the performance and uncertainties of models to choose the most suitable biomass function for a specific application.

4.1 Materials and methods

Biomass is frequently estimated using models, given limitations in direct measurement methods (Gonçalves et al., 2019a). Biomass modelling is challenging as it has to accommodate a range of variables such as differences in forest species and stand structure (Gonçalves, 2017, 2018). In this study, two approaches were used. The first approach involved modelling forest biomass for five species (*Quercus rotundifolia*, *Quercus suber*, *Pinus pinea*, *Pinus pinaster* and *Eucalyptus* spp.) in pure and mixed stands combining data from forest inventory and high spatial resolution satellite image data. The second approach involved evaluation of potential forest residues for energy.

The first approach included the development of several aboveground biomass models. Biomass models at the area level for *Quercus rotundifolia*, *Pinus pinea*, *Pinus pinaster* and *Eucalyptus* spp. were developed for pure stands (i.e. composed mainly by one species) using biomass estimated at tree level (with published allometric equations and forest inventory data) and upscaled to a per area unit (per hectare) as a dependent variable. Crown horizontal projection area or crown cover or vegetation indices, resulting from de vegetation mask derived from high spatial resolution satellite images data, were explanatory variables (Gonçalves et al., 2017a,b; Macedo et al., 2018a; Sousa et al., 2015). From available mathematical models, single or multiple linear regression and random forest models.

4.2 Results

Overall, the performance of all models was good with errors within an acceptable error threshold. The model for *Quercus rotundifolia* (with crown horizontal projection as an explanatory variable) (Sousa et al., 2015) had better predictive ability than that with one vegetation index (NDVI, SR or EVI) as an explanatory variables (Macedo et al., 2018b). This might be due to *Quercus rotundifolia* forest systems having an irregular spatial distribution with many isolated trees isolated whilst others were in groups, with crowns in the vertical profile centred in one layer with a low number of trees per unit area. The model developed for *Pinus pinaster*, with data from high spatial resolution satellite images, enabled the evaluation of the biomass dynamics in time (for 2004, 2007 and 2011) as well as the effect of forest fires and the pinewood nematode pest on aboveground biomass (Gonçalves et al., 2017a).

Biomass models for pure and mixed stands of *Quercus rotundifolia*, *Quercus suber* and *Pinus pinea* were also developed with crown horizontal projection or crown cover (either total or per species) derived from high spatial resolution satellite images data and dummy variables for composition as explanatory variables (Gonçalves et al., 2019b; Sousa et al., 2017). For both approaches, the best performing models included total crown horizontal projection (Sousa et al., 2017) or crown cover (Gonçalves et al., 2019b) and dummy variables for composition, the latter denoting the differences in stand structure.

Aboveground biomass models for *Quercus rotundifolia*, *Quercus suber* and *Pinus pinea* were fitted random forests and explanatory variables derived from high spatial resolution satellite images data, including spectral bands, vegetation indices, texture indices and crown horizontal projection, calculated with and without the vegetation mask (Lourenço et al., 2021). The texture index

Grey-Level Co-occurrence Matrix attained the highest relative performance, followed by the vegetation indices and the spectral bands. This relation was stronger for data calculated with the vegetation mask than without it. This could be due to the irregularity of the spatial distribution pattern of the trees (isolated and in groups) and the low number of trees per unit area (Lourenço et al., 2021).

The second approach focused on assessment of biomass for energy. Potential forest residues were assessed using a geographical information system (GIS) framework. The evaluation of the spatial distribution of forest residues from silvicultural practices (cut, thinning and pruning) was based on a collection area close to a bioenergy facility (Mesquita et al., 2018; Rocha et al., 2020). Other studies included the availability of forest residues of *Pinus pinaster* for bioenergy (Malico et al., 2017); a time series developed by Gonçalves et al. (2017a), including the effect of forest tree growth, fire and pinewood nematode on the potential amount of forest residues for energy on a per year basis (Malico et al., 2017), and the potential availability of biomass for energy of *Quercus rotundifolia* and *Quercus suber* stands, using the biomass function of Gonçalves et al. (2019b) and estimating the forest residues for residential heating, considering several scenarios of pruning intensity and different conversion technologies (Malico et al., 2021).

Funding: This work was funded by National Funds through FCT (Foundation for Science and Technology) under the Project UIDB/05183/2020.

5 Assessing the carbon footprint of horticultural crops

The world population has reached 8 billion people (UN, 2022). Feeding this population is degrading terrestrial and aquatic ecosystems and contributing to climate change (Poore and Nemecek, 2018). Agricultural production contributes between 30% and 40% of total anthropogenic GHG emissions contributing to climate change (Parajuli et al., 2018; Poore and Nemecek, 2018). The reduction of carbon emissions and GHG emissions at a global level was established in the Paris agreement (2016) and reaffirmed at COP26 (2021). The EU has also proposed to include in its 'Fit for 55 legislative package' (a set of measures to achieve carbon neutrality) an increase in carbon sequestration through changes in land use, forestry and agriculture, aiming to achieve climate neutrality by 2035 (EU Climate Action, 2022). Portugal is obligated to implement measures to achieve 'Land use' climate neutrality by 2035.

The carbon footprint (CF) is a commonly used indicator to quantify CE and GHG emissions (Chen et al., 2021). One of the most commonly used methods to assess CF is life cycle assessment (LCA) which translates production data into environmental impacts (Poore and Nemecek, 2018) (Fig. 11). LCA encompasses

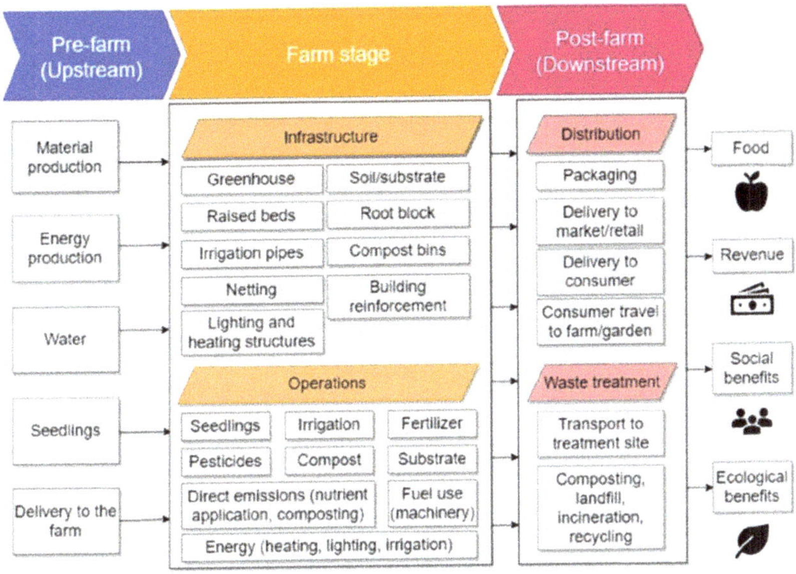

Figure 11 Example of an LCA and the points to be included in the assessment. They are considered the most important procedures of the production cycle (extracted from Dorr et al., 2021).

activities during production (nutrition, irrigation, etc.), which have been divided in the case of horticulture into six steps (Parajuli et al., 2018):

1 Nursery sowing;
2 Land preparation and sowing/planting in the final location;
3 Initial phase of plant development;
4 Full vegetative growth;
5 Crop senescence; and
6 Harvest and disposal of crop residues.

LCA also takes into account the pre-production (production of fertilizers, energy, etc.), and post-production stages (packaging, storage, retail display, waste, etc.) (Dorr et al., 2021). The study in this section reviews current LCA research in horticulture, and addresses the following questions: What is the average carbon footprint in potato, onion, carrot, melon, and watermelon crop production in Portugal? Which agricultural operations have the highest and lowest carbon footprint? How can the production system be optimized?

The authors found 122 mainly European LCA studies, with tomato the most studied crop (Clune et al., 2016; Dorr et al., 2021), as shown in Table 7 (Chen et al., 2021). It can be difficult to compare data (from the same product or

Table 7 Summary of LCA studies found on horticultural crops and GWP average global values (kg CO_2-eq/kg product)

Culture		LCA (no. of studies)	GWP (average global value)
Open air	Tomato	**19**	0.46
	Potato	16	0.20
	Lettuce	15	0.37
	Strawberry	15	0.65
	Carrot	10	0.22
	Onion	7	**0.18**
	Broccoli	6	0.70
	Asparagus	5	0.83
	Pumpkin	4	0.33
	Cauliflower + broccoli	4	0.35
	Watermelon	2	0.32
Heated greenhouse	Tomato	13	2.69
	Lettuce	3	**3.15**
	Strawberry	3	2.56

Values taken from (Clune et al., 2016).

Table 8 Emissions with the greatest impact identified in the LCA of horticultural crops

GHG Sources	%
Irrigation energy	54
Soil N emission *(after fertilization)*	17
Energy used in post-harvest *(storage)*	11
Applied fertilizer *(input)*	10
Fuel *(used in crop operations)*	8

Values taken from (Clune et al, 2016)

different products), since study methodologies are different which influences the results obtained (Clune et al., 2016; Dorr et al., 2021). The global warming potential (GWP), expressed in kg CO_2-eq/kg product, is the key metric used in LCA studies (Clune et al., 2016).

Current LCA studies (Table 7) suggest that outdoor horticultural production contributes for a lower carbon footprint than indoor horticulture. Onions cultivation has the lowest CO2-eq emission (GWP = 0.18 kg CO_2-eq/kg product). Lettuce production in a heated greenhouse has the highest impact (GWP = 3.15 kg CO_2-eq/kg product) (Clune et al., 2016). These studies suggest that producing a crop out of season and/or in a region with unfavourable climatic conditions dramatically increases its environmental impact. This is evident e.g. when comparing outdoor tomato production (GWP = 0.46 kg CO_2-eq/kg product) to tomato production in a heated greenhouse (GWP = 2.69 kg CO_2-eq/kg product).

The activities in horticultural production that contribute most to emissions are energy used for irrigation and direct emissions caused by the use of synthetic fertilizers (Clune et al., 2016), as shown in Table 8. Inputs such as fertilizers and pesticides are considered the most significant contributors to the carbon footprint of horticulture. GHG emissions mainly occur at three critical points in the crop cycle: during the soil preparation phase, the phase of full crop growth (due to fertilizer application and energy used for irrigation) and the harvest/post-harvest phase (due to fuel consumption and energy used in processing and storage).

5.1 Materials and methods

The calculation of the carbon footprint in this study was performed according to the methodology developed by the IPCC (2006), considering two main types of GHG sources:

- Direct sources: fuel consumption, application of synthetic fertilizers and land-use change (when native vegetation is removed for other purposes); and
- Indirect sources: volatilization, leaching and runoff of CO2-eq (considering biochemical processes and climatic conditions).

The life cycle of each crop begins with soil tillage and ends with harvest. In this study transportation of products to storage and subsequent processing were not considered. To calculate GHG emissions, data were collected between 2019 and 2021 for each production cycle:

- Soil management and fertilization: types of fertilizers used and the quantity applied (Kg ha^{-1});
- Fuel: date of refuelling, vehicle details, quantity refuelled, vehicle work location, and specific mode of operation;
- Harvest: date of harvest, quantity harvested, yield % sold and yield % discarded; and
- Irrigation: irrigation volume (L ha^{-1}), irrigation frequency and energy consumption.

This allowed the calculation of both direct and indirect GHG emissions.

LCA studies were conducted on experimental farms situated in the Ribatejo Region in Portugal. According to the Köppen classification, the climate corresponds to subtype Csa, characterized by hot, dry summers with mostly clear skies. Winters are cool, with moderate precipitation, overcast skies, occasional frost and fog. Throughout the year, temperatures range between 5°C and 33°C, rarely falling below 0°C or exceeding 40°C. Insolation (the amount of solar radiation reaching the region) is high and can exceed 3000 annual hours. Soils are predominantly sandy with gentle slopes, with low fertility and low organic matter content (<1%), slightly acidic pH (between 6 and 7) and water average usable capacity of 0.08 $m^3 m^{-3}$. Sprinkler and drip irrigation systems using groundwater sources are used predominantly. The horticultural crops for this study includes potatoes, onions, carrots, watermelons and melons. Different varieties were sown/planted for each crop, as indicated in Fig. 12.

5.2 Results

In general, average GHG emissions obtained for different crops were quite close, with watermelon having the lowest average GHG emissions at 0.04 kg CO2-eq/kg product (Fig. 13). Potatoes and onions showed identical average GHG emissions of 0.06 kg CO2-eq/kg product. However, looking at the maximum values, potatoes have the highest recorded value at 0.14 kg CO2-eq/kg product. Potatoes also exhibit the highest variation in results. Carrots have the lowest variation, with an average value of 0.05 kg CO2-eq/kg product, a minimum value of 0.03 kg CO2-eq/kg product and a maximum value of 0.06 kg CO2-eq/kg product. The largest source of GHG emissions came from fuels used in various crop operations (Fig. 14a), accounting for 58% of the total emissions, followed by fertilizer application with 31%. Looking at the distribution of fuel consumption per operation, harvesting operations account for 41% of total fuel consumption. Soil tillage, also has a significant impact, accounting for 23% (Fig. 14b).

As this analysis shows, the variation in carbon footprints is primarily dependent on three main factors: yield, fuel consumption and soil management (tillage and fertilizer application). This suggests that strategies to mitigate carbon footprint and adapt to climate change in fruits and vegetables production involves the implementation of:

- Soil and water conservation practices;
- Integrated pest and disease management;
- Cultivation management with increased rotation, re-scheduling of planting and harvesting;

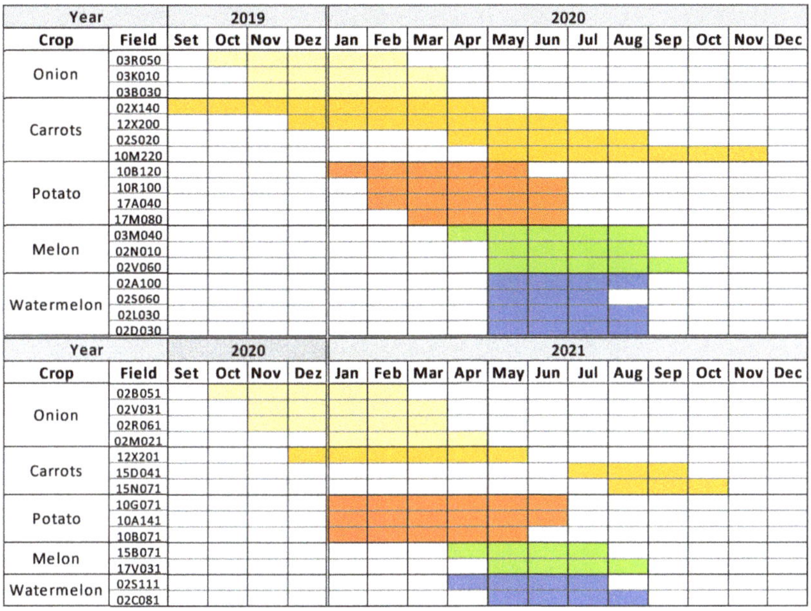

Figure 12 Season and duration of the crops Life Cycle of the cultures, between 2019 and 2021.

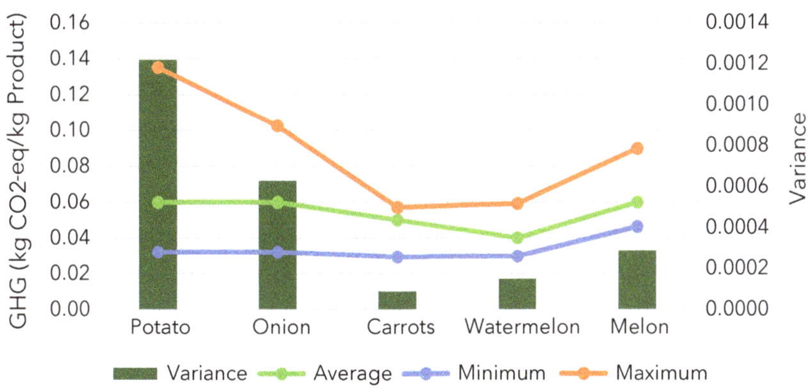

Figure 13 GHGE ($kg_{CO2\text{-}eq}/kg_{product}$) average, minimum and maximum values obtained in the analysed crops in 2020 and 2021.

- Utilization of species and varieties tolerant to climate stress and diseases;
- More efficient irrigation practices; and
- Use of renewable energy sources.

Other interventions include increasing green fertilization.

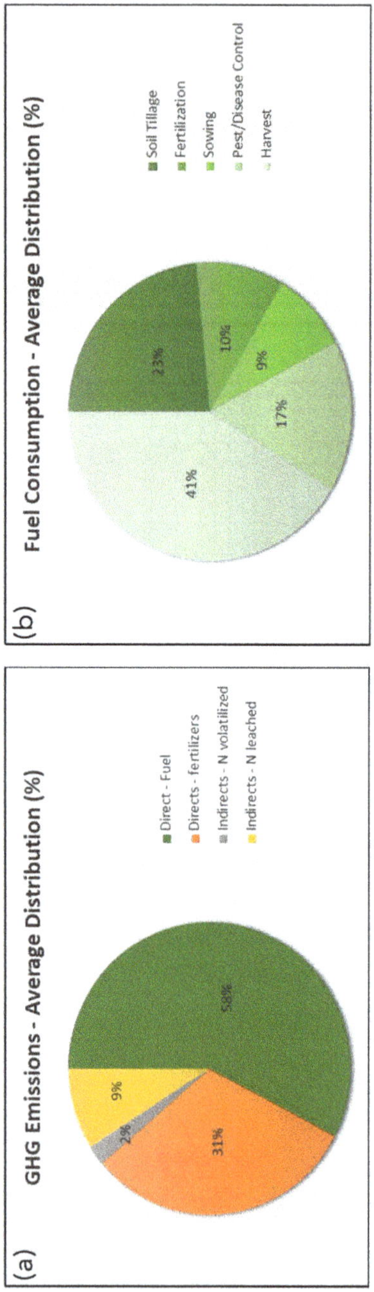

Figure 14 (a) Average distribution of GHGE by Sources in horticultural crops, (b) Average distribution of Fuel Consumption by cultural operations carried out in the horticultural production

6 Assessment of animal welfare using the Aurora app

Animal welfare, along with environmental and food safety issues, is currently considered one of the greatest challenges in livestock farming. The environmental conditions to which animals are exposed throughout their production process can have a significant impact on animal welfare, particularly in terms of maintaining thermal balance with the surrounding environment. This impact is typically greater in intensive production systems characterized by accelerated production rates and high animal densities.

The Aurora application was developed by Oliveira Júnior et al. (2019) as a mobile application to create maps that show the spatial variability of thermal comfort indices within animal housing. The application was specifically developed for the Android operating system, starting from version 8.0. The development of the application was carried out using the Android Studio 2.3.3 integrated development environment, along with the Java object-oriented programming language. Maps were created using the inverse distance weighted (IDW) interpolation method. The MPAndroidChart library (Jahoda, 2018) was also used, specifically implementing the ScatterChart mode. The size of the chart area is adapted according to the capacity of the mobile device screen. The width of the map (width) is configured using the getSize method of the Display class, ensuring that the map is proportional to the width of the smartphone screen.

The process of interpolating and plotting data on the map requires prior configuration of the arrangement of known points. Once this configuration is done, the application is able to determine the size of the area and, consequently,

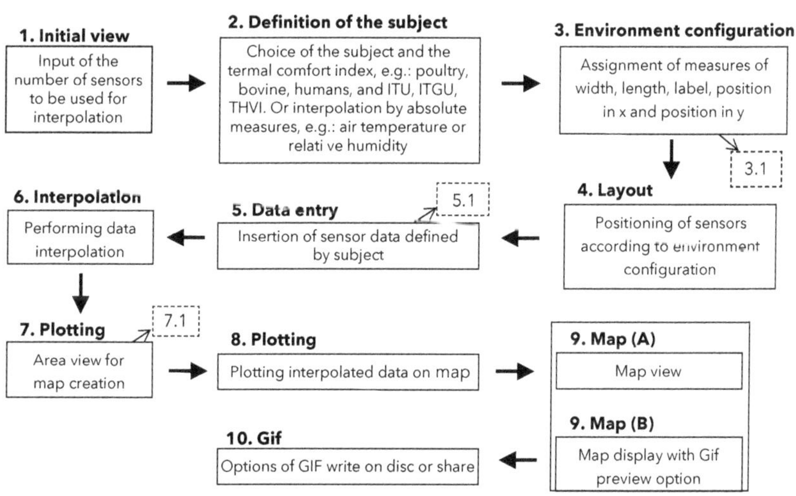

Figure 15 Aurora data entry and output flowchart. Source: Adapted from: Oliveira Júnior et al. (2019) and Oliveira Júnior (2019).

create a grid of points. It also calculates the distances between known and estimated points. Based on meteorological parameters provided by the user, the application performs the interpolation and then creates a map that represents the thermal condition of the environment, both in terms of thermal comfort indices and air temperature measurements. The air temperature and thermal comfort index maps are created using a gradient of 64 colours, where lower values are represented in blue and higher values in red. Figures 15 and 16 illustrate the flowcharts that demonstrate the sequence of input, data processing and output in the application.

The Aurora application was developed on the Google Play platform for Android smartphones. It is available for download in both Portuguese and English versions (https://play.google.com/store/apps/details?id=com.olvappaurora). The application has been validated (Oliveira Júnior et al., 2019) and proved to be a useful tool for evaluating microclimate conditions inside animal facilities. The

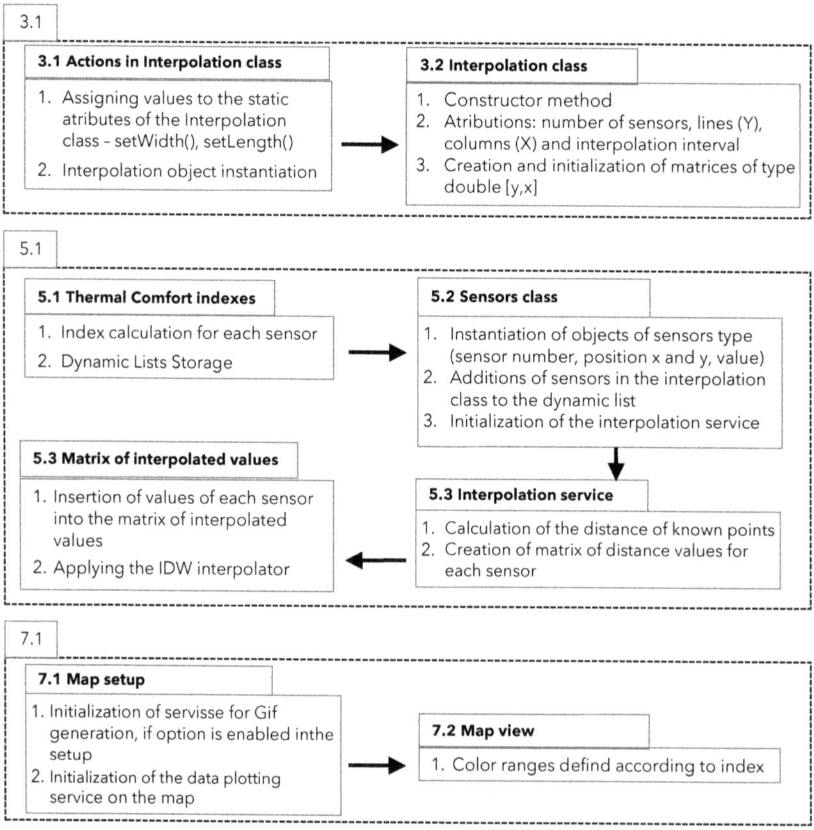

Figure 16 Detailing of the internal processes of the Aurora application. Source: Adapted from: Oliveira Júnior et al. (2019) and Oliveira Júnior (2019).

application is limited in application to larger areas given the process of plotting the interpolated data on a graph, which requires a high memory consumption of mobile devices. Given such limitations, this application should be considered only as an auxiliary tool for studying the thermal conditions of animals.

7 Conclusion and future trends

This chapter has presented five case studies focused on the application of precision agriculture/smart agriculture in different production systems. These help to contribute to more resilient and sustainable crop, livestock and forest production, by reducing environmental impacts, GHG emissions and production costs.

The first study case shows the results of a long-term study to calibrate a Grassmaster II capacitance probe to estimate pasture productivity in the Mediterranean *Montado* ecosystem. The second presents a cost-effective methodology to identify yield zones within an olive grove, contributing for the development of a decision support system (DSS) for precision olive farming. The third case study presents an overview of the different remote sensing sensors, data and methodologies mostly used to estimate forest biomass in Alentejo, south of Portugal. A fourth case study reviews the carbon footprint of horticultural crops to identify which crop practices contribute to emissions and how to optimize the production system in order to reduce the carbon footprint. The final case study deals with a precision livestock farming (PLF) mobile app to identify indicators related to animal welfare.

Initiatives such as the EU's Green Deal, Farm to Fork Strategy and the new Common Agricultural Policy (CAP) constitute a roadmap of policies, actions, incentives and management measures to promote the efficient use of resources in agriculture, reduce its carbon footprint, restore and preserve biodiversity and rural landscapes. To achieve these objectives requires innovative and sustainable solutions supported by disruptive technologies (IoT, big data, artificial intelligence, etc.) and based on a holistic approach, in which the promotion of the circular economy, the expansion of conservation (or regenerative) agriculture practices, decentralized food chains, citizens participation and comprehensive, equal and inclusive digitalization in agriculture will play a key role.

Research in the use of proximal and remote sensing data for agricultural and forestry systems is likely to continue evolving in response to emerging technologies, data analytics techniques and the growing need for sustainable land management practices. This will lead to development of more advanced and cost-effective remote sensing technologies, including improvements in satellite sensors, drones (Unmanned Aerial Vehicles - UAVs) and proximal sensors. It will also lead to better methods to integrate and analyse data from multiple sources such as machine learning and artificial intelligence algorithms. These researches can help automate the process of identifying and classifying different tree and crop species, and monitoring changes over time, allowing optimization of the resources use such as water, fertilizers and pesticides based on real-time data, thereby improving efficiency and sustainability. Decision support systems can be developed to help policymakers and farmers make informed choices concerning production practices. The future will be characterized by integration of cutting-edge technologies, data analysis methods and a strong focus on sustainability and climate resilience. This research will continue to play a crucial role in promoting sustainable land management and food security. Find more information about these topics, here https://www.med.uevora.pt/med-publications/.

8 References

Al-Bakri, J. T. and Taylor, J. C. (2003). Application of NOAA AVHRR for monitoring vegetation conditions and biomass in Jordan, *Journal of Arid Environments* 54(3):579–593. https://doi.org/10.1006/jare.2002.1081.

Cauduro, G. F., Carvalho, P., Barbosa, C. M. P., Lunardi, R., Pilau, A., Freitas, F. K. de. and Silva, J. L. Sd. (2006). A comparison of methods used to determine forage mass in Italian ryegras pasture (*Lolium Multiflorum* Lam.), *Ciência Rural* 36(5):1617–1623.

Chen, R., Zhang, R. and Han, H. (2021). Where has carbon footprint research gone?, *Ecological Indicators* 120:106882. https://doi.org/10.1016/j.ecolind.2020.106882.

Cicore, P., Serrano, J., Shahidian, S., Sousa, A., Costa, J. L. and da Silva, J. R. (2016). Assessment of the spatial variability in tall wheatgrass forage using LANDSAT 8 satellite imagery to delineate potential management zones, *Environmental Monitoring and Assessment* 188(9):513.

Clune, S., Crossin, E. and Verghese, K. (2016). Systematic review of greenhouse gas emissions for different fresh food categories, *Journal of Cleaner Production* 140(part 2):766–783. https://doi.org/10.1016/j.jclepro.2016.04.082.

Directorate-General for Climate Action. (2022). Land use, forestry and agriculture. Disponível em https://climate.ec.europa.eu/eu-action/european-green-deal/delivering-european-green-deal/land-use-forestry-and-agriculture_en.

Díez, C. M., Moral, J., Cabello, D., Morello, P., Rallo, L. and Barranco, D. (2016). Cultivar and tree density as key factors in the long-term performance of super high-density olive orchards, *Frontiers in Plant Science* 7:1226.

Dorr, E., Goldstein, B., Horvath, A., Aubry, C. and Gabrielle, B. (2021). Environmental impacts and resource use of urban agriculture: A systematic review and meta-analysis, *Environmental Research Letters* 16(9):093002. https://doi.org/10.1088/1748-9326/ac1a39.

Efe Serrano, J. (2006). *Pastures in Alentejo: Technical Basis for Characterization, Grazing and Improvement*. Évora: Universidade de Évora, ICAM, pp. 165–178.

Fernández-Manso, O., Fernández-Manso, A. and Quintano, C. (2014). Estimation of aboveground biomass in Mediterranean forests by statistical modelling of ASTER fraction images, *I.J. of Applied Earth Observation and Geoinformation* 31:45–56. https://doi.org/10.1016/j.jag.2014.03.005.

Fonseca, A. D. and Fernandes, J. C. (2004). *Detecção REMOTA*. Lisboa: LIDEL. EDIÇÕES TÉCNICAS.

Ghasemi, N., Sahebi, M. R. and Mohammadzadeh, A. (2011). A review on biomass estimation methods using synthetic aperture radar data, *Int. J. Geomat. Geoscience* 1(4):776–788.

Gonçalves, A. C. (2017). Multi-species stand classification: definition and perspectives. In Chakravarty, S. and Shukla, G. (Eds.), *Forest Ecology and Conservation*. InTech, pp. 3–23. https://doi.org/10.5772/67662.

Gonçalves, A. C. (2018). Effects of forest stand structure in biomass and carbon. In Shukla, G. and Chakravarty, S. (Eds.), *Forest Biomass and Carbon*. Rijeka: InTech, pp. 1–21. https://doi.org/10.5772/intechopen.76004.

Gonçalves, A. C., Malico, I. and Sousa, A. M. O. (2021). Energy production from forest biomass: An overview. In Gonçalves, A. C., Sousa, A. and Malico, I. (Eds.), *Forest Biomass - From Trees to Energy*. London: IntechOpen, pp. 1–23. https://doi.org/10.5772/intechopen.93361.

Gonçalves, A. C., Sousa, A. M. O. and Mesquita, P. G. (2017a). Estimation and dynamics of above ground biomass with very high resolution satellite images in Pinus pinaster stands, *Biomass and Bioenergy* 106:146–154. https://doi.org/10.1016/j.biombioe.2017.08.026.

Gonçalves, A. C., Sousa, A. M. O. and Silva, J. R. M. (2017b). *Pinus pinea* above ground biomass estimation with very high spatial resolution satellite images. In: Carrasquinho, I., Correia, A. C. and Mutke, S. (Eds.). Mediterranean pine nuts from forests and plantations. Zaragoza: CIHEAM. p. 49–54 (Options Méditerranéennes: Série A.Séminaires Méditerranéens; n. 122).

Gonçalves, A. C., Malico, I. and Sousa, A. M. O. (2019a). Solid biomass from forest trees to energy: A review. In Jacob-Lopes, E. and Queiroz Zepka, L. (Eds.), *Renewable Resources and Biorefineries*. IntechOpen. https://doi.org/10.5772/intechopen.79303.

Gonçalves, A. C., Sousa, A. M. O. and Mesquita, P. (2019b). Functions for aboveground biomass estimation derived from satellite images data in Mediterranean agroforestry systems, *Agroforestry Systems* 93(4):1485–1500. https://doi.org/10.1007/s10457-018-0252-4.

Hanna, M. M., Steyn-Ross, D. A. and Steyn-Ross, M. (1999). Estimating biomass for New Zealand pasture using optical remote sensing techniques, *Geocarto International* 14(3):89–94.

Haralick, R. M., Shanmugam, K. and Dinstein, I. (1973). Textural features for image classification, *IEEE Transactions on Systems, Man and Cybernetics* 3:610–621.

Hirata, M. (2000). Quantifying spatial heterogeneity in herbage mass and consumption in pastures, *Journal of Range Management* 53(3):315–321.

IPCC. (2006). *IPCC 2006 Revised Good Practice Guidelines for Greenhouse Gas Inventories*. Tokyo: Intergovernmental Panel on Climate Change (IPCC), Institute for Global Environmental Strategies. Disponível em http://www.ipcc-nggip.iges.or.jp/public/2006gl/vol4.html. Accessed 11 July 2011.

Jahoda, P. (2018). MPAndroidChart. https://github.com/PhilJay/MPAndroidChart.

Jamieson, P. D., Porter, J. R. and Wilson, D. R. (1991). A test of the computer simulation model ARCWHEAT1 on wheat crops grown in New Zealand, *Field Crops Research* 27(4):337–350.

Jolliffe, I. T. (2002). *Principal Component Analysis* (2nd edn.). New York, NY: Springer-Verlag.

Júnior, O. (2019). A. J. Sistema móvel para análise de conforto térmico de animais, utilizando método interpolador IDW. PhD Thesis. Botucatu: Universidade Estadual Paulista (UNESP), 112p.

Karl, M. G. and Nicholson, R. A. (1987). Evaluation of the forage disk method in mixed grass rangeland in Kansas, *Journal of Range Management* 40(5):467–471.

LaRue, E. A., Atkins, J. W., Dahlin, K., Fahey, R., Fei, S., Gough, C. and Hardiman, B. S. (2018). Linking Landsat to terrestrial LiDAR: vegetation metrics of forest greenness are correlated with canopy structural complexity, *International Journal of Applied Earth Observation and Geoinformation* 73:420–427.

Leboeuf, A., Beaudoin, A., Fournier, R., Guindon, L., Luther, J. and Lambert, M. (2007). A shadow fraction method for mapping biomass of northern boreal black spruce forests using QuickBird imagery, *Remote Sensing of Environment* 110(4):488–500. https://doi.org/10.1016/j.rse.2006.05.025.

Li, X., Du, H., Mao, F., Zhou, G., Chen, L., Xing, L., Fan, W., Xu, X., Liu, Y., Cui, L., Li, Y., Zhu, D. and Liu, T. (2018). Estimating bamboo forest aboveground biomass using EnKF-assimilated MODIS LAI spatiotemporal data and machine learning algorithms, *Agricultural and Forest Meteorology* 256–257:445–457. https://doi.org/10.1016/j.agrformet.2018.04.002.

Lourenço, P., Godinho, S., Sousa, A. and Gonçalves, A. C. (2021). Estimating tree aboveground biomass using multispectral satellite-based data in Mediterranean agroforestry system using random forest algorithm, *Remote Sensing Applications: Society and Environment* 23:100560. https://doi.org/10.1016/j.rsase.2021.100560.

Macedo, F. L., Sousa, A. M. O., Gonçalves, A. C., Silva, H. R. and Rodrigues, R. A. F. (2018a). Função alométrica de biomassa com imagens de satélite de alta resolução espacial, *Ciência Florestal* 28(3):960. https://doi.org/10.5902/1980509833368.

Macedo, F. L., Sousa, A. M. O., Gonçalves, A. C., Marques da Silva, J. R., Mesquita, P. A. and Rodrigues, R. A. F. (2018b). Above-ground biomass estimation for Quercus rotundifolia using vegetation indices derived from high spatial resolution satellite images, *European Journal of Remote Sensing* 51(1):932-944. https://doi.org/10.1080/22797254.2018.1521250.

Malico, I., Gonçalves, A. C. and Sousa, A. (2017). Assessment of the availability of forest biomass for biofuels production in Southwestern Portugal, *Defect and Diffusion Forum* 371:121–127. https://doi.org/10.4028/www.scientific.net/DDF.371.121.

Malico, I., Gonçalves, A. C. and Sousa, A. M. O. (2021). Evergreen oak biomass residues for firewood. In Gonçalves, A. C., Sousa, A. M. O. and Malico, I. (Eds.), *Forest Biomass - From Trees to Energy*. London: IntechOpen, pp. 87–103. https://doi.org/10.5772/intechopen.95417.

Mesquita, P., Nepomuceno Pereira, R., Malico, I., Gonçalves, A. C. and SOUSA, A. M. O. (2018). GIS based analysis of potential forest residues for energy in Alentejo, Portugal Graz. In:Proceeding of International Sustainable Energy Conference 2018, Graz, Áustria, 3–5 October 2018, p.664–671.

Murphy, W. M., Silman, J. P. and Mena Barreto, A. D. (1995). A comparison of quadrat, capacitance meter, HFRO sward stick, and rising plate meter for estimating herbage mass in a smooth-stalked meadowgrass-dominant white clover sward, *Grass and Forage Science* 50:452–455.

Newnham, G. J., Armston, J. D., Calders, K., Disney, M. I., Lovell, J. L., Schaaf, C. B., Strahler, A. H. and Danson, F. M. (2015). Terrestrial laser scanning for plot-scale forest measurement, *Current Forestry Reports* 1(4):239–251.

Ogura, S. and Hirata, M. (2001). Two-dimensional monitoring of spatial distribution of herbage mass in a bahia grass (Paspalum notatum Flugge) pasture grazed with cattle, *Grassland Science* 47:453–459.

Oliveira Júnior, A. J., Souza, S. R. L., Dal Pai, E., Rodrigues, B. T. and Souza, V. C. A. (2019). Mobile application for analysis of spatial variability of thermal comfort indexes of animals and people, using IDW interpolation, *Computers and Electronics in Agriculture* 157:98–101.

Poore, J. and Nemecek, T. (2018). Reducing food's environmental impacts through producers and consumers, *Science* 360(6392):987–992. https://www.science.org/doi/10.1126/science.aaq0216.

Pullanagari, R. R., Yule, I. J., Tuohy, M. P., Hedley, M. J., Dynes, R. A. and King, W. M. (2013). Proximal sensing of the seasonal variability of pasture nutritive value using multispectral radiometry, *Grass and Forage Science* 68(1):110–119.

Quegan, S., Le Toan, T., Yu, J. J., Ribbes, F. and Floury, N. (2000). Multitemporal ERS SAR analysis applied to forest mapping Geosci, *IEEE Transactions on Geoscience and Remote Sensing* 38(2):741–753.

Reese, G. A., Bayn, R. L. and West, N. E. (1980). Evaluation of double-sampling estimators of subalpine herbage production, *Journal of Range Management* 33(4):300–306.

Rocha, J. T., Malico, I., Gonçalves, A. C. and Sousa, A. M. O. (2020). Análise do potencial de biomassa residual no Algarve, Portugal, baseada em SIG, *Rev. Ciênc. Madeira: RCM* 11(1):42-52. https://doi.org/10.12953/2177-6830/rcm.v11n1p42-52.

Rouse, J. W, Haas, R. H., Schell, J. A. and Deering, D. W. (1973). Monitoring Vegetation Systems in the Great Plains with ERTS (Earth Resources Technology Satellite). *Proceedings of 3rd Earth Resources Technology Satellite Symposium* 351: 309–317.

Sanderson, M. A., Rotz, C. A., Fultz, S. W. and Rayburn, E. B. (2001). Estimating forage mass with a commercial capacitance meter, rising plate meter, and pasture ruler, *Agronomy Journal* 93(6):1281–1286.

Schaefer, M. T. and Lamb, D. W. (2016). A combination of plant NDVI and LiDAR measurements improve the estimation of pasture biomass in Tall Fescue (Festuca arundinacea var. Fletcher), *Remote Sensing* 8(2):1–10.

Schellberg, J., Hill, M. J., Gerhards, R., Rothmund, M. and Braun, M. (2008). Precision agriculture on grassland: Applications, perspectives and constraints, *European Journal of Agronomy* 29(2–3):59–71.

Serrano, J. M., Peça, J. O., Marques da Silva, J. and Shahidian, S. (2011). Calibration of a capacitance probe for measurement and mapping of dry matter yield in Mediterranean pastures, *Precision Agriculture* 12(6):860–875.

Serrano, J., Shahidian, S., Moral, F. and Marques da Silva, J. M. (2019). Long-term evaluation of the Grassmaster II probe to estimate productivity of pastures. In Stafford, J. V. (Ed.), Precision Agriculture'19, Proceedings of the 12th European Conference on Precision Agriculture (ECPA2019). Wageningen: Wageningen Academic Publishers, pp. 631–637.

Silveira, D. L. (2022). Otimização da gestão de uma parcela de olival em sebe partindo de imagens de deteção remota – um caso prático. Dissertação de Mestrado em

Tecnologias em Agricultura de Precisão. Universidade de Évora - Escola de Ciências e Tecnologia; Universidade Nova de Lisboa - Faculdade de Ciências e Tecnologias.

Sousa, A. M. O., Gonçalves, A. C. and Silva, J. R. M. (2017). Above-ground biomass estimation with high spatial resolution satellite images. In Tumuluru, J. S. (Ed.), *Biomass Volume Estimation and Valorization for Energy*. InTech. https://doi.org/10.5772/65665.

Sousa, A. M. O., Gonçalves, A. C., Mesquita, P. and Marques da Silva, J. R. (2015). Biomass estimation with high resolution satellite images: A case study of Quercus rotundifolia, *ISPRS Journal of Photogrammetry and Remote Sensing* 101:69–79. https://doi.org/10.1016/j.isprsjprs.2014.12.004.

Townsend, P. A. (2002). Estimating forest structure in wetlands using multitemporal SAR, *Remote Sensing of Environment* 79(2–3):288–304.

United Nations (2022). Total population - Both sexes. consultado a 2022/12/06.

Wang, M., Cao, W., Guan, Q., Wu, G., Jiang, C., Yan, Y. and Su, X. (2018). Potential of texture metrics derived from high-resolution PLEIADES satellite data for quantifying aboveground carbon of Kandelia candel mangrove forests in Southeast China, *Wetlands Ecology and Management* 26(5):789–803. https://doi.org/10.1007/s11273-018-9610-2.

Watzlawick, L. F., Koehler, H. S. and Kirchner, F. F. (2006). Estimativa de biomassa e carbono em plantios de Pinus taeda L. utilizando imagens do satélite IKONOS II, *Ciência e Natura* 28:45–60.

Vickery, P. J. and Nicol, G. R. (1982). An improved electronic capacitance meter for estimating pasture yield: Construction details and performance tests (pp. 1–22). Australian Animal Research Laboratories technical paper no. 9. Armidale: CSIRO.

Virkajarvi, P. (1999). Comparison of three indirect methods for prediction of herbage mass on timothymeadow fescue pastures, *Soil and Plant Science* 49:75–81.

Zanine, A. M., Santos, E. M. and Ferreira (2006). Main methodologies for evaluation of pastures, *Revista Electrónica de Veterinaria, REDVET* 7:1–13.

Zhang, Q., He, H. S., Liang, Y., Hawbaker, T. J., Henne, P. D., Liu, J., Huang, S., Wu, Z. and Huang, C. (2018). Integrating forest inventory data and MODIS data to map species-level biomass in Chinese boreal forests, *Canadian Journal of Forest Research* 48(5):461–479. https://doi.org/10.1139/cjfr-2017-0346.

Zheng, D., Rademacher, J., Chen, J., Crow, T., Bresee, M., Le Moine, J. and Ryu, S. (2004). Estimating aboveground biomass using Landsat 7 ETM+ data across a managed landscape in northern Wisconsin, USA, *Remote Sensing of Environment* 93(3):402–411. https://doi.org/10.1016/j.rse.2004.08.008.

Zhu, X. and Liu, D. (2015). Improving forest aboveground biomass estimation using seasonal Landsat NDVI time-series, *ISPRS Journal of Photogrammetry and Remote Sensing* 102:222–231. https://doi.org/10.1016/j.isprsjprs.2014.08.014.

Chapter 4

Agri Semantics: developments to improve data interoperability to support farm information management and decision support systems in agriculture

Saba Noor, Jade Bokma and Bart Pardon, Ghent University, Belgium; Gerdien van Schaik, Utrecht University, The Netherlands; and Miel Hostens, Cornell University, USA

1 Introduction

1.1 What is Agri Semantics?

In farm animal health management systems (FAHMS), Agri Semantics is part of smart agri systems and is crucial to improving data management and analysis. Semantic web technology is the extension of the World Wide Web and aims to make web content more machine understandable. Ontologies are the fundamental part of developing and implementing a semantic web (Roussaki et al., 2023). Ontologies are the formal and explicit descriptions of data models that define concepts, relationships, and properties within a specific application domain. With the help of Semantic web technologies, Agri Semantics (Khatoon and Ahmed, 2022) enables the structured and semantically enriched representation of farm animal health-related data, leading to several benefits such as improving knowledge representation, data integration, data interoperability, data analysis, and empowering decision support systems that help veterinarians and farmers in making informed decisions related to disease management. By using semantic web technologies, Agri Semantics contributes to more effective disease prevention, early detection, and improved animal health outcomes in the agricultural sector (Niloofar et al., 2021).

http://dx.doi.org/10.19103/AS.2023.0132.05

Data-intensive farming (Bullock et al., 2019) refers to the practice of collecting, analyzing, and utilizing agricultural data for the management and optimization of farming practices. It involves the collection of data from various sources, such as sensors, weather stations, and farm management software to get useful information and make data-driven decisions in agricultural operations. Farmers are encouraged to share their farm-related data in exchange for rewards such as financial rewards, access to specific services, and personalized recommendations, leading to a comprehensive data ecosystem. Through data-driven insights and collaboration, data-intensive farming improves productivity, sustainability, and knowledge sharing in agriculture. On the other hand, precision livestock farming (PLF) is an advanced technology to automate real-time monitoring of farm animal health data, productivity, environmental impact, and welfare (Bahlo et al., 2019). Due to the advancement in this technology, a large amount of data has become available, but the data availability (Villeneuve et al., 2019), confidentiality, accessibility, interoperability, and complexity of such heterogeneous data remain challenging (Hermans et al., 2018). The usefulness of PLF can be increased if the data from different sources are processed together to give more meaningful and unexplored insights (Symeonaki et al., 2022). Although PLF appeals to young people and helps agriculture, it can harm farmers and animals if not tailored to their needs and skills (Fote et al., 2020). From the farmer's point of view, PLF techniques erode their observation skills and increase dependency on tools.

The term 'farm animal health' describes the mental and physical well-being of farm animals. It incorporates various aspects, such as disease prevention, veterinary care, nutrition, and overall welfare. On the other hand, farm management systems incorporate the overall practices and strategies employed by farmers' points of view to raise and care for their animals. While farm animal health and farm animal management systems are interconnected, they are not synonymous. The management system plays a significant role in influencing animal health outcomes. For example, the housing system chosen can impact animal welfare, disease transmission, and stress levels. In this context, the goal of the EU-funded project called DECIDE (European Horizon, 2020 project https://decideproject.eu/) is to develop decision-support tools for improved control of endemic infectious diseases in livestock and aquaculture. DECIDE focuses on respiratory and gastrointestinal syndromes in swine, poultry, cattle, and growth reduction in salmonids. The project involves identifying stakeholder needs, accessing relevant data, determining disease burden and control costs, implementing early warning systems, and understanding disease impact. The early warning system within DECIDE aims to provide timely information to stakeholders, such as farmers and veterinarians, enabling them to make informed decisions on disease-mitigating actions. By detecting signals early, stakeholders can take proactive measures to prevent or control disease spread and minimize the impact on animal health and welfare (van Schaik et al., 2023).

1.2 Data challenges in traditional farm animal health management systems and the importance of Agri Semantics?

In the context of PLF, traditional farm animal information systems pose significant challenges to data collection, cleaning, integration, and analysis (Rosa, 2021). One of the key data challenges is the lack of standardization of data representation. The collected data comes from various sources such as laboratories, veterinary clinics, and research institutes and often follow a different format and lack a unified structure. This heterogeneity and lack of standardization make it difficult to effectively utilize the data for decision-making. Such systems lack extensibility and are not stable to communicate in heterogeneous systems (Niloofar et al., 2020). Another challenge is the limited interoperability between different disease management systems and data sources. In traditional systems, data exchange and collaboration among stakeholders are hindered by incompatible formats and are difficult to share, therefore limiting the ability to collaborate and exchange data among stakeholders. Additionally, traditional information management systems (Monteiro et al., 2021) face difficulties to ensure data quality and completeness. Outdated, incomplete, and inaccurate data compromise the reliability of disease surveillance, diagnosis, and decision-making. The lack of quality control mechanisms may lead to inconsistencies and errors in the collected data. The lack of advanced analytical capabilities for knowledge discovery and decision support (Newton et al., 2020) is another significant data challenge. This hinders the identification of patterns and trends in disease occurrence, correlation, and transmission. Thus, it limits the ability to develop effective information management strategies and timely interventions.

Addressing these data challenges is crucial for improving the effectiveness of traditional information management systems. The adoption of modern approaches (Newton et al., 2020), such as ontology-based systems supported by Agri Semantics, plays an important role in addressing these data challenges by providing a framework for structuring and harmonizing data in a meaningful way. By implementing ontologies, Agri Semantics enables the creation of a common understanding and representation of agricultural concepts, facilitating data integration across different sources, providing interoperability, and enabling advanced data analysis (Murali and Anouncia, 2022). It allows for the seamless exchange and sharing of data, enabling collaborative efforts in farm animal disease management. By addressing these challenges, stakeholders can enhance disease surveillance, early detection, and control efforts, ultimately leading to improved animal health outcomes and promoting effective prevention and more efficient disease management practices (Hu et al., 2020).

1.3 Existing animal livestock ontologies and techniques

In the context of PLF, numerous ontology-based approaches were proposed, such as AGROVOC agriculture thesaurus (Rajbhandari and Keizer, 2012), Animal Trait Ontology for Livestock (ATOL) (Golik et al., 2012), Infectious Disease Ontology (IDO) (Cowell and Smith, 2010), Animal Disease Ontology (ADO) (Malhotra et al., 2014), and Animal Health Surveillance Ontology (AHSO) (Dórea et al., 2019). However, these approaches do not provide a dedicated ontology focusing on disease and pathogen identification in farm animals. Moreover, these ontologies do not provide the level of granularity for our cattle barometer (https://decideproject.eu/barometer/) use case.

Ontologies in farm animal health management systems have drawbacks, including a lack of user knowledge. A practical approach to address these limitations is to develop a cattle barometer use case framework. By developing a species-specific framework, we can help guide the implementation and incorporation of ontologies in the context of cattle health management. This will help us to monitor the implementation process, gather feedback from users, and evaluate the impact of the framework on cattle health and management outcomes. This step will help identify challenges and areas for improvement.

1.4 Problem statement

In the field of FAHMS the transmission of infectious diseases poses a significant challenge, leading to threats to animal health, welfare, and reduced productivity. The datasets related to transmission and contagious diseases often exhibit variations in format, semantics, syntax, and structure. To address the heterogeneity of data, we need to design a cost-effective, accessible, reusable, and lightweight ontology-driven knowledge-based domain-specific framework for farm animal health management systems. This chapter focuses on a cattle barometer use case scenario, which serves as a representative example for the application of the proposed ontology-driven knowledge-based framework in farm animal health management systems. The cattle barometer use case scenario serves as practical and illustrative to represent the effectiveness and applicability of the proposed framework in real-world farm animal health management systems that follows the FAIR (Findable, Accessible, Interoperable, Reusable) guideline principles (van Schaik et al., 2023).

The key contribution of this research include the following:

1. Establishing a standardized framework on FAIR guideline principles to tradeoff between data requirements, data accessibility, availability, data connectivity, and value generation;
2. Developing a species-specific ontology called Livestock Health Ontology (LHO) in which knowledge graphs are enriched by semantic knowledge coming from domain knowledge;
3. Generating Resource Descriptive Framework (RDF) data from the raw data of the cattle use case scenario, leveraging the advantages of RDF for enhanced data sharing, integration, and querying capabilities;
4. Mapping of cleaned RDF data to the species-specific LHO ontology for semantic interoperability; and
5. Utilizing a data visualization tool for visualizing and reporting farm animal disease management data to gain meaningful insight, enhancing data interpretation, and decision-making capabilities.

1.5 Organization of the chapter

The remaining sections of the chapters are organized as follows: In Section 2, we provide background knowledge and a literature review. In Section 3, we present the methodology in detail. In Section 4, we describe a case study integrating and analyzing data from several animal health data sources. In Section 5, we conclude the paper and discuss future research goals and challenges.

2 Background knowledge and literature review

2.1 Ontology-based approaches

In the field of PLF, numerous ontological sources have been developed such as the AGROVOC agriculture thesaurus (Rajbhandari and Keizer, 2012), the most popular one managed by the Food and Agriculture Organization of the United Nations (FAO) which aligns with the FAIR principles (Meyer et al., 2021). It is a linked open data for agriculture consisting of over 3200 concepts and available in 20 different languages. It serves as valuable resource for capturing general concepts regarding agriculture and animal health. However, AGROVOC does not contain a dedicated ontology focusing on disease and pathogen identification in farm animals (Gangadharan and Gupta, 2020).

Other smaller ontologies are IDO (Cowell and Smith, 2010), Coronavirus Infectious Disease Ontology (CIDO), and VIDO named Virus Infectious Disease Ontology (Babcock et al., 2021). These ontologies are the structural representation of infectious diseases and rely on domain-specific extension for example tuberculosis and coronaviruses of various species like cattle, pigs, poultry, and fish. ATOL (Golik et al., 2012) is another ontology that models a multi-species ontology for animal phenotype traits. It focuses on animal

Table 1 Comparative analysis of ontology-based versus non-ontology-based animal health management systems

Ontology-based			Non-ontology based		
Ontology name	Remarks	Reference	Method name	Remarks	Reference
AGROVOC	Constantly updated	https://agrovoc.fao.org/browse/agrovoc/en/index/B	RISKSUR EVA	Non-ontological base but by using EVA tool may in cooperate the elements of ontologies	https://fp7-risksur.eu/results/tools
IDO	Published 2014 and last updated on 08-03-2022	https://bioportal.bioontology.org/ontologies/IDO	EpiTools	Calculating various epidemiological measures	https://epitools.ausvet.com.au/
CIDO	Extension of IDO, updated on August 3, 2020	https://bioportal.bioontology.org/ontologies/IDO-COVID-19	Surveillance Evaluation Tool (SET)	Measure the performance of animal disease surveillance	https://extranet.who.int/sph/index.php/evaluation-action-fao-surveillance-evaluation-tool-set
VIDO	Extension of IDO, updates on November 27, 2022	https://bioportal.bioontology.org/ontologies/VIDO	--------	---------	--------------
ATOL	Published on 2014 and updated May 11, 2020	https://bioportal.bioontology.org/ontologies/ATOL	---------	----------	-----------------
ADO	Last uploaded: November 14, 2018	https://agroportal.lirmm.fr/ontologies/ANDO	--------	---------	--------------
AHSO	Last update 2018	Link available on Bioportal and Github: https://bioportal.bioontology.org/ontologies/AHSO/?p=summary https://github.com/SVA-SE/AHSO	--------	---------	--------------

Published by Burleigh Dodds Science Publishing Limited, 2024.

products like meat, milk quality, and quantity, reproductivity, and welfare and was last updated in 2014. Similarly, ADO (Malhotra et al., 2014) focuses on representing the knowledge of animal diseases specifically focus 500 pairs of genetic disorders and infectious and metabolic diseases of different animal species. It focuses on the concepts regarding disease treatment, pathogens, diagnostic tests, symptoms, and progression. AHSO (Dórea et al., 2019) focuses on the relation and concepts of animal health surveillance. It provides structural information about disease reporting, monitoring, and surveillance activities. However, these ontologies do not provide the level of granularity that we need for our cattle barometer use case. The conclusion is that the existing ontologies cannot be reused as they are and need to be extended or modified to align with the desired level of detail and compatibility of relations and concepts with the design ontology. An overview of ontology-based versus non-ontology-based animal health management systems is given in Table 1.

2.2 Non-ontology-based approaches (traditional approaches)

Traditional farm animal information systems like Epitools are used for epidemiological analysis, but their usability for non-veterinarians has been questioned (Cocca and Lindberg, 2019), although valuable for decision-making in animal health. Similarly, the Surveillance Evaluation Tool (SET) (Vasconcelos Gioia et al., 2021) is developed by WOAH (World Organisation for Animal Health) to measure the performance and use of animal disease surveillance. To use this tool effectively, it may require a good understanding of surveillance methodologies and principles. RiskSUR (Peyre et al., 2019) is a tool that provides a framework for designing, implementing, analyzing, and reporting surveillance systems. The aforementioned systems have limitations to incorporate advanced modeling technologies consequently providing less accessibility, interoperability, reusability, and availability (Bhat and Huang, 2021). Furthermore, such type of systems are not farmer friendly and requires user expertise to handle and manage.

3 Methodology

3.1 System model

Figure 1 depicts the data flow of a generic non-ontology based versus ontology-driven knowledge-based framework of Farm Animal Data Management ODKFADM. Non-ontology-based systems are systems that do not use ontologies. Instead they rely on relational databases or other data management and statistical techniques. In such systems, the process of converting raw data into meaningful statistical results is repetitive, time-consuming, and tedious work (Ren et al., 2020). Every time when a new dataset comes in, it requires

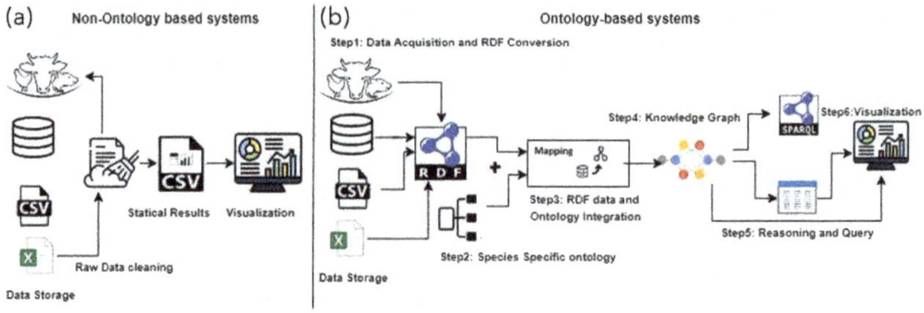

Figure 1 (a) Non-ontology based versus (b) ontology-driven Knowledge-based framework of Farm Animal Data Management (ODKFADM).

rewriting data cleaning code because the naming convention, format, and structure of each dataset are different and require specific adjustments to the data cleaning and processing steps. Therefore, non-ontology-based systems lack standardized semantic structures and reasoning that involve inferring new knowledge from existing datasets and relationships, limiting the system's overall flexibility, adaptability, reusability, and integrity.

However, in ODKFADM the data stored in heterogeneous resources are exposed as a knowledge graph and enriched with semantic knowledge coming from domain knowledge. This ontology mapping results in knowledge graph (Chen et al., 2020), the heterogeneous device data can be expressed in terms of ontologies. Ontologies provide an explicitly structured and standardized formal representation of different concepts, classes, properties, and relationships within the domain of animal farm management. Once the ontology is established, the concepts and relationships remain unchanged nevertheless of the specific dataset being processed. In formal ontology-based approaches (Saha, 2021) that support knowledge graphs, frequent code modification is limited. Therefore, an ontology-driven approach can be reusable promoting interoperability, flexibility, and integrity. It also facilitates querying and reasoning over the data at the ontological level.

4 Case study

4.1 Dataset and test bed

We evaluate the ODKFADM on data focusing on cattle. It involves collection of raw data from different animal health service labs in Europe from different countries. These datasets include information about farms, their geolocation, breed type (beef, dairy, or mixed), diagnostic tests, pathogen identification, and pathogen results for specific bacteria such as *Histophilus somni* (HS), *Mannheimia haemolytica* (MH), *Pasteurella multocida* (PM), *Mycoplasma bovis*

(MB), and viruses such as bovine coronavirus (BCV) and bovine respiratory syncytial virus (BRSV).

The aim is to demonstrate the usefulness of the ODKFADM for analyzing and understanding cattle-related raw data by using the R, Python and Pyspark framework to enhance data integration and advance semantic reasoning capabilities. Moreover, it also improves decisions on disease control and increases animal health and welfare. To achieve this, the ODKFADM framework follows a multistep approach. In addition, the R, Pyspark, and pandas notebook code is uploaded on GitHub, making it publicly available for use. This allows other researchers, data scientists, and stakeholders to access and utilize the code for their own cattle-related data analysis projects. By sharing these resources, we can foster collaboration and knowledge exchange, leading to further advancements in disease control, animal health, and welfare. The code is available at decide-project-eu/LiveStockHealthOntologyWp1 (github.com), and the ontology can be accessed on BioPortal at https://bioportal.bioontology.org/ontologies/LHO version 1.0.0.

4.2 Step 1: data acquisition and resource descriptive frameworks conversions

The code for step 1, as illustrated in Fig. 2, is available on GitHub. In this step, we read the raw data from heterogeneous resources using pandas, Pyspark, and R data frames and then convert the data frames into RDF (resource descriptive frameworks) format. Basically, the RDF (Ali et al., 2022) format is a way to represent the data in an explicit standardized format. To perform these conversions, we utilize the RDFLib library in Python and the RDF packages in R. During RDF conversion, we implement anonymization techniques to protect sensitive information. For this purpose, personal identifiers such as sample numbers, farm identification, and file numbers were anonymized by using the cryptographic SHA 256 hash function. SHA256 hash function ensured that the sensitive information could not be linked back to a specific entity or individual while preserving identifications.

4.3 Step 2: species-specific Livestock Health Ontology

Ontology is the formal and explicit description of a data model within an application domain in which relationships occur between individuals, making it easy to reuse and share knowledge. LHO is an extension of the DECIDE ontology. LHO follows OBO Foundry guidelines and contains information related to the health and well-being of livestock animal species such as cattle. Furthermore, it represents the explicit structured and standardized framework for specific domain knowledge and data integration of livestock species. In the

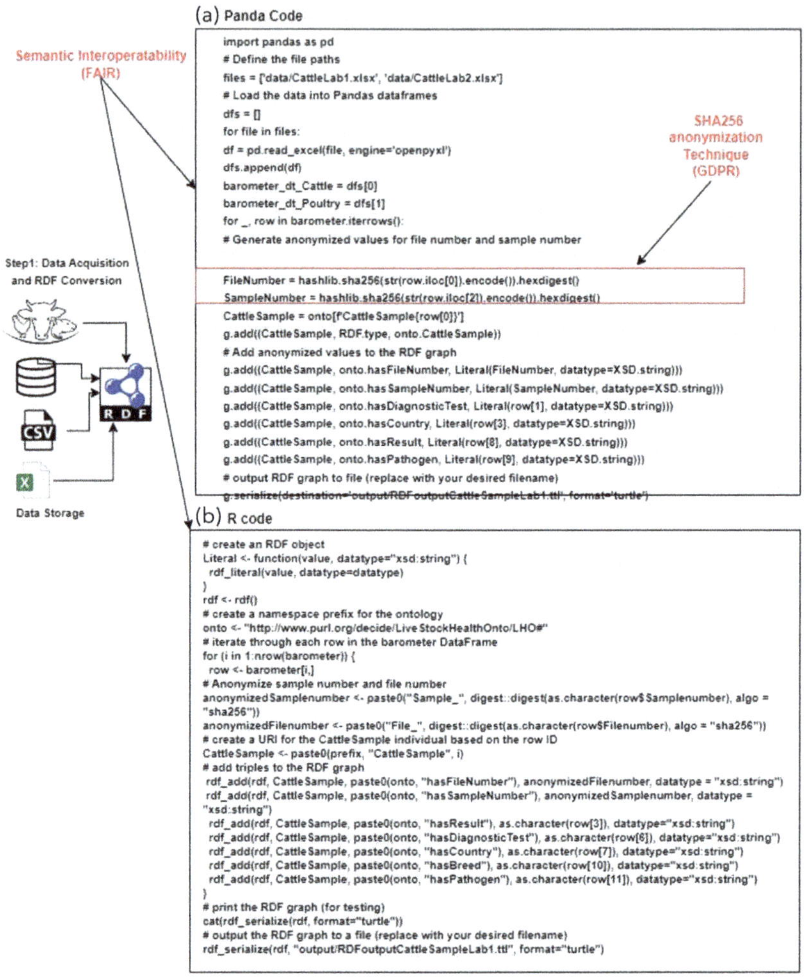

Figure 2 Data acquisition and RDF conversions.

development process, knowledge can be acquired from domain experts, and the initial prototype is designed to facilitate reasoning specifically for cattle-related data. Furthermore, LHO contains information on anonymized farm identification, samples, geolocation, breed, diagnostic tests, and pathogen results for bacteria such as HS, MH, PM, and viral pathogen like BCV along with respiratory culture results. This information is represented in the form of concepts of classes, subclasses, and their relationship in the form of object and data properties. The LHO is uploaded on Biopoartal and can be accessed at https://bioportal.bioontology.org/ontologies/LHO. By making LHO available on Bioportal, we aim to promote its use and encourage community feedback in the field of farm animal health management.

4.3.1 Formal representation of the Livestock Health Ontology

Ontologies belong to the specific domain environment that contains a semantic level, in terms of vocabulary and their relationship (Gómez-Pérez and Corcho, 2002). The LHO model can be formally represented as:

LHO = (C, P, A, I)

Where LHO stands for Livestock Health Ontology model, C is the concepts and instances conceptualization, P is the properties and their relationships, A represents the rules or axioms, and I represent the model interpretation. In LHO, the object properties (P) and relationships (Re) can be formally represented by the following symbols:

P = (D: Data type; O: Object; T: Transitive; F: Functional)

Re = (≡: equivalence; ⊂: subsumed; ∩: Disjoint)

The domain of property is formally represented as follows:

[∀b; b: p (a; b) → Domain (a)]; p∈ P; Domain ∈ C

The range of property is formally represented as follows:

[∀b; b: p (a; b) → Range (b)]; p ∈ P; Range ∈ C

The interpretation of the ontology model is represented as LHO. Classes and properties possess many Re such as subsuming, equivalence, and disjoint relationships. The symbol ≡ shows the equivalence of classes; the symbol ∩ shows conceptually disjoint classes and is symmetric, reflexive, and transitive. Some examples of these relationships in the LHO ontology model are

CattleSample ∩ PoultrySample ∩ PigSample ⊂ *LiveStockHealthOnto*

In the LHO model, all (⊂) subsume relations are asymmetric, irreflexive, and transitive. However, ∩ is conceptually disjoint, and it never violates its property of being transitive (relationships can be inherited), symmetric (relationships are bidirectional), and reflexive (entities can have a relation with themselves). To make the LHO model consistent, it is necessary to fulfill all these properties. Some examples of these hierarchal relationships in the LHO ontology model are

LiveStockHealthOnto ⊂ *LiveStockHealthOntology*

CattleSample ⊂ *LiveStockHealthOnto*

4.3.2 Define classes and subclasses of Livestock Health Ontology

In LHO, *LiveStockHealthOntology* is represented as the top-level concept. *SampleType, SampleNumber, FarmIdentification, Breed, Province, PostalCode, Country, DiagnosticTest, PathogentIdentification, PathogenResults*, etc. are the subclasses of the *LiveStockHealthOnto* class, as shown in Fig. 3.

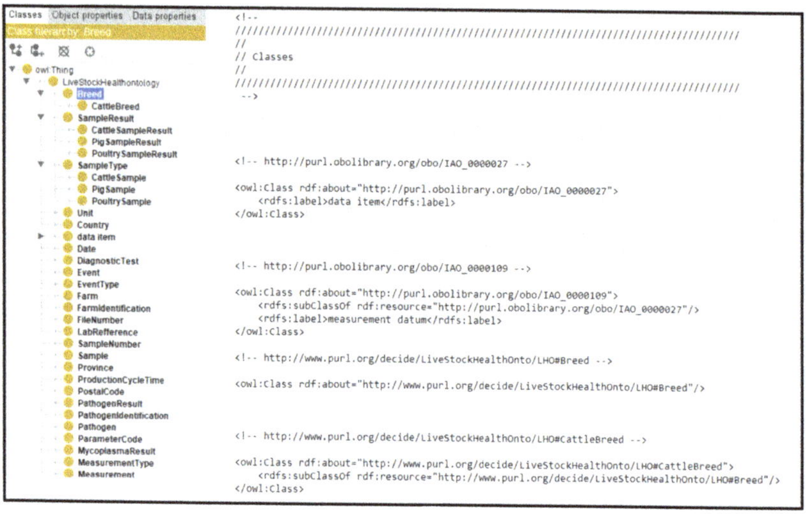

Figure 3 Define classes and subclasses of LHO.

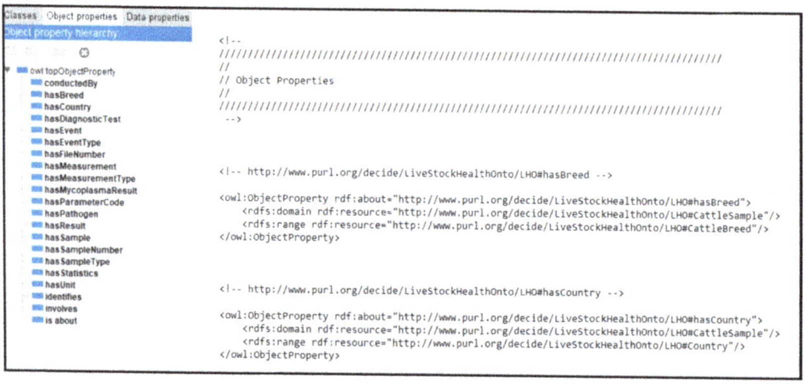

Figure 4 Define object properties of LHO.

4.3.3 Define object properties of Livestock Health Ontology

Figure 4 shows the relationships of object properties in Ontology web language (Owl) and RDF such as *hasBreed*, *hasCountry*, *hasPathogenIdentification*, and *hasPathogenresults*.

4.3.4 Define data properties of Livestock Health Ontology

The proposed ontology also defines data properties to represent the relationship between individuals and data literals. Annotation properties can be used for describing metadata properties. Figure 5 shows the data properties and its domain and range of particular data properties using semantic web technologies Owl and RDF schemas.

4.3.5 Define Livestock Health Ontology axioms

LHO axioms represent the relationship between classes, instances, and attributes. For example, Fig. 6 shows class axioms describing the semantic structure and meaning such as Owl: disjoint class, rdfs: Subclassof, rdf: Description, and rdf: type. After defining these axioms, we can provide the disjoint and equivalence relationships among classes.

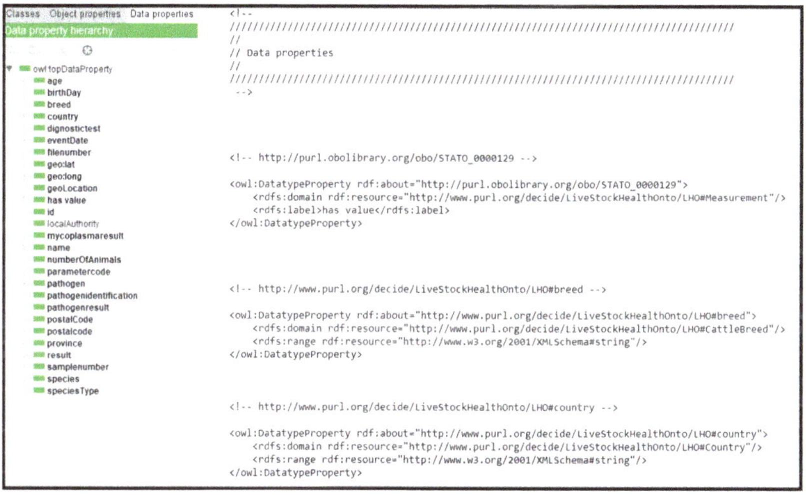

Figure 5 Define data properties of LHO.

```
    <!--
    ///////////////////////////////////////////////////////////////////////////////////////
    //
    // General axioms
    //
    ///////////////////////////////////////////////////////////////////////////////////////
     -->

    <rdf:Description>
        <rdf:type rdf:resource="http://www.w3.org/2002/07/owl#AllDisjointClasses"/>
        <owl:members rdf:parseType="Collection">
            <rdf:Description rdf:about="http://www.purl.org/decide/LiveStockHealthOnto/LHO#CattleSample"/>
            <rdf:Description rdf:about="http://www.purl.org/decide/LiveStockHealthOnto/LHO#PigSample"/>
            <rdf:Description rdf:about="http://www.purl.org/decide/LiveStockHealthOnto/LHO#PoultrySample"/>
        </owl:members>
    </rdf:Description>
    <rdf:Description>
        <rdf:type rdf:resource="http://www.w3.org/2002/07/owl#AllDisjointClasses"/>
        <owl:members rdf:parseType="Collection">
            <rdf:Description rdf:about="http://www.purl.org/decide/LiveStockHealthOnto/LHO#CattleSampleResult"/>
            <rdf:Description rdf:about="http://www.purl.org/decide/LiveStockHealthOnto/LHO#PigSampleResult"/>
            <rdf:Description rdf:about="http://www.purl.org/decide/LiveStockHealthOnto/LHO#PoultrySampleResult"/>
        </owl:members>
    </rdf:Description>
</rdf:RDF>
```

Figure 6 Define axioms of LHO.

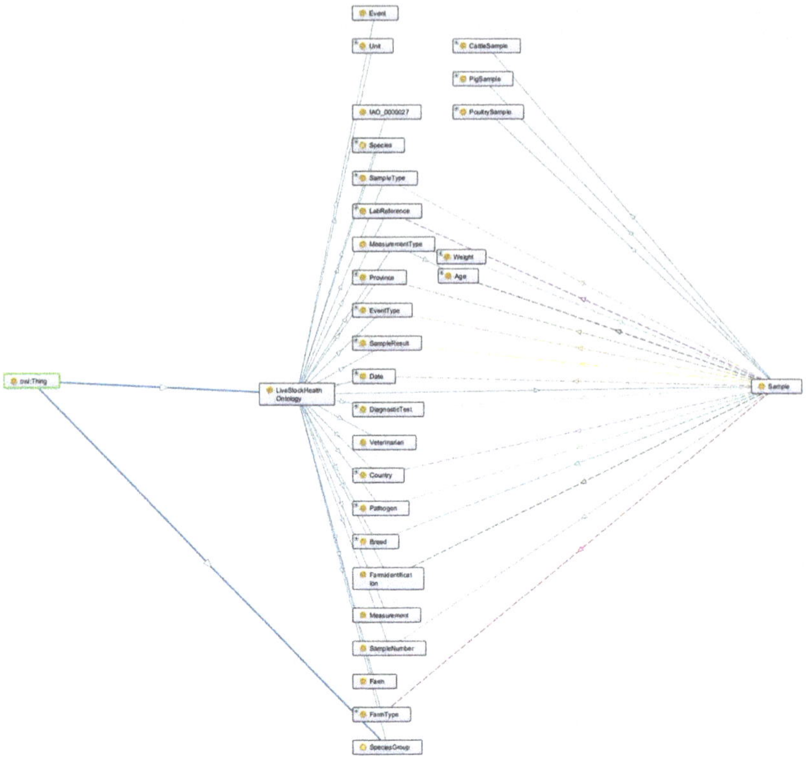

Figure 7 Graphical representation of LHO.

4.3.6 Graphical representation of ontology

Figure 7 shows the graphical representation of ontology in Onto Graph representing classes, subclasses, and their relationships.

4.4 Step 3: Resource Descriptive Framework data and ontology integration (mapping)

With the help of RDF data representation, we map the RDF data explained in step 1 with LHO in step 2. This mapping process (Sahoo et al., 2009) enriches the data with semantic meanings, thus enabling advanced reasoning and querying as shown in Fig. 8.

4.5 Step 4: knowledge graph (ontology update)

The knowledge graph is a well-unified, structured, and integrated knowledge repository. When the RDF data are integrated with the ontology, it may require updating existing ontology consequently contributing to the development

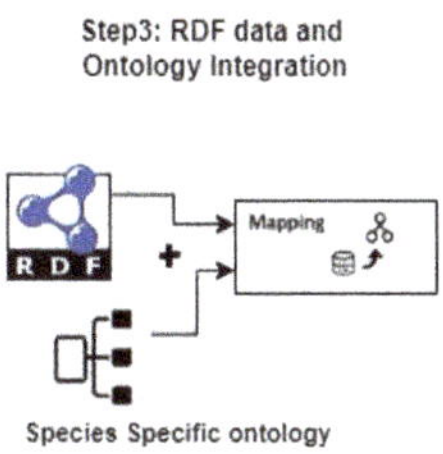


```
# Step 3: Load the RDF data and ontology into a Panda DataFrame:
# Define the original path
path_to_RDF = "output/RDFoutputCattleSampleLab1.ttl"
   # Create a new graph
   g = Graph()
   # Parse the RDF file in Turtle format
   g.parse(path_to_RDF, format='ttl')

   # Parse the ontology file in OWL format and add it to the graph
   path_to_ontology = "Ontology/LivestockHealthOnto1.0.owl"
   g.parse(path_to_ontology, format="xml")

#Use RDFS or OWL reasoning to infer additional knowledge
g.bind('rdfs', RDFS)
g.bind('owl', OWL)
g.bind('onto', Namespace("http://www.purl.org/decide/LivestockHealthOnto"))
```

Figure 8 RDF data and ontology integration (mapping).

	FileNumber	DiagnosticTest	SampleNumber	Breed	LabReference	SampleType	PostalCode	Province	PathogenIdentification	PathogenResult	MycoplasmaResult
0	381ab40916b42502cd14b9de7201d798d31c35255ccd2a...	Culture	854116431e19bdb205ca2a76204cda6aba4069bb500bdc...	Beef	1	Autopsy	9800	nan	Histophilus somni	POS++	POS++
1	381ab40916b42502cd14b9de7201d798d31c35255ccd2a	Culture	854116431e19bdb205ca2a76204cda6aba4069bb500bdc	Beef	1	Autopsy	9800	nan	Histophilus somni	POS++	POS++
2	381ab40916b42502cd14b9de7201d798d31c35255ccd2a...	Culture	854116431e19bdb205ca2a76204cda6aba4069bb500bdc...	Beef	1	Autopsy	9800	nan	Mannheimia haemolytica	POS++	POS++
3	381ab40916b42502cd14b9de7201d798d31c35255ccd2a...	Culture	854116431e19bdb205ca2a76204cda6aba4069bb500bdc...	Beef	1	Autopsy	9800	nan	Mannheimia haemolytica	POS++	POS++
4	381ab40916b42502cd14b9de7201d798d31c35255ccd2a...	Culture	854116431e19bdb205ca2a76204cda6aba4069bb500bdc...	Beef	1	Autopsy	9800	nan	Mycoplasma bovis	POS++	POS++

Figure 9 Query 1: filtering positive PathogenResults and MycoplasmaResults associated with CattleSample.

of a knowledge graph. Such updating could be done by adding new classes, relationships, and properties concepts for a better understanding of integrated data, thus resulting in a knowledge graph that enhances more advanced reasoning and querying capabilities (Hogan et al., 2021).

4.6 Step 5: reasoning and query

To explore the knowledge graph, we have used reasoning and SPARQL queries (Kollia et al., 2011). The following section elaborates on different SPARQL queries on LHO.

4.6.1 Query 1: filtering positive PathogenResults and MycoplasmaResults associated with CattleSample

Figure 9 shows the query that filters positive PathogenResult and MycoplasmaResult which is associated with CattleSample-related data. For each CattleSample, the results show the information of file number, diagnostic tests, breed, sample type, sample number, lab reference, geolocation, and pathogen information.

```
PREFIX decide: <http://www.purl.org/decide#>
PREFIX LHO: <https://www.purl.org/decide/LiveStockHealthOnto/LHO#>
```

Published by Burleigh Dodds Science Publishing Limited, 2024.

```
SELECT ?FileNumber ?DiagnosticTest ?SampleNumber ?Breed ?LabReference ?SampleType ?Result ?Pathogen ?PostalCode ?Province ?PathogenIdentification ?PathogenResult ?MycoplasmaResult
WHERE {
?CattleSample LHO:hasFileNumber ?FileNumber .
?CattleSample LHO:hasDiagnosticTest ?DiagnosticTest .
?CattleSample LHO:hasSampleNumber ?SampleNumber .
?CattleSample LHO:hasBreed ?Breed .
?CattleSample LHO:hasLabReference ?LabReference .
?CattleSample LHO:hasSampleType ?SampleType .
?CattleSample LHO:hasResult ?Result .
?CattleSample LHO:hasPathogen ?Pathogen .
?CattleSample LHO:hasPostalCode ?PostalCode .
?CattleSample LHO:hasProvince ?Province .
?CattleSample LHO:hasPathogenIdentification ?PathogenIdentification .
?CattleSample LHO:hasPathogenResult ?PathogenResult .
FILTER (?PathogenResult = "POS++")
?CattleSample LHO:hasMicoplasmaResult ?MycoplasmaResult .
FILTER (?MycoplasmaResult = "POS++")
```

4.6.2 Query 2

Breed has three types: beef, meat, or mixed. The SPARQL query 2 retrieves the results based on Breed begin Beef as shown in Fig. 10

```
PREFIX decide: <http://www.purl.org/decide#>
PREFIX LHO: <https://www.purl.org/decide/LiveStockHealthOnto/LHO#>
SELECT ?FileNumber ?DiagnosticTest ?SampleNumber ?Breed ?LabReference ?SampleType ?Result ?Pathogen ?PostalCode ?Province ?PathogenIdentification ?PathogenResult ?MycoplasmaResult

WHERE {
?CattleSample LHO:hasFileNumber ?FileNumber .
?CattleSample LHO:hasDiagnosticTest ?DiagnosticTest .
?CattleSample LHO:hasSampleNumber ?SampleNumber .
?CattleSample LHO:hasBreed ?Breed .
```

	FileNumber	DiagnosticTest	SampleNumber	Breed	LabReference	SampleType	Result	Pathogen	PostalCode	Province	PathogenIdentification	PathogenResult	MycoplasmaResult
0	952fe97b89661dc65d3c1bae7e87dcac5f10cd96cc92b2...	PCR	22a38575c079e307143b9fcfb252d27eefea042fbbc956...	Beef	1	BAL	0	AD Corona Ag (PCR)	8700	West Flanders	nan	nan	nan
1	952fe97b89661dc65d3c1bae7e87dcac5f10cd96cc92b2...	PCR	6e116f0e1b58751c4ad0d8144618e77fe8f1792437291	Beef	1	BAL	0	AD Corona Ag (PCR)	8700	West Flanders	nan	nan	nan
2	952fe97b89661dc65d3c1bae7e87dcac5f10cd96cc92b2...	PCR	f81c46f81ec28d93f7485884ec185b48241202fd2430b3 ...	Beef	1	BAL	0	AD Corona Ag (PCR)	8700	West Flanders	nan	nan	nan
3	952fe97b89661dc65d3c1bae7e87dcac5f10cd96cc92b2	PCR	22a38575c079e307143b9fcfb252d27eefea042fbbc956...	Beef	1	BAL	1	AD Corona Ag (PCR)	8700	West Flanders	nan	nan	nan
4	952fe97b89661dc65d3c1bae7e87dcac5f10cd96cc92b2...	PCR	6e116f0e1b58751c4ad0d8144618e77fe8f1792437291...	Beef	1	BAL	1	AD Corona Ag (PCR)	8700	West Flanders	nan	nan	nan

Figure 10 Query 2: filtering the results based on the breed being 'Beef' associated with CattleSample.

```
FILTER (?Breed = "Beef")
?CattleSample LHO:hasLabReference ?LabReference .
?CattleSample LHO:hasSampleType ?SampleType .
?CattleSample LHOhasResult ?Result .
?CattleSample LHO:hasPathogen ?Pathogen .
?CattleSample LHO:hasPostalCode ?PostalCode .
?CattleSample LHO:hasProvince ?Province .
?CattleSample LHO:hasPathogenIdentification ?PathogenIdentification .
?CattleSample LHO:hasPathogenResult ?PathogenResult .
?CattleSample LHO:hasMicoplasmaResult ?MycoplasmaResult .
}
```

4.7 Step 6: visualization and analysis

On the basis of our knowledge graph complexity, tools, and techniques, we choose the tool Tableau for visualization. Tableau provides meaningful insights to explore the knowledge graph. For this, we need a working ODBC connection to a Virtuoso Instance and ODBC or JDBC Compliant version of Tableau or Tableau Server (Heward-Mills, 2017). We can create a visualization specifically tailored for the cattle barometer use case, focusing on respiratory infectious diseases. This visualization, known as the cattle barometer, is a powerful visualization tool designed to provide meaningful insights and understanding

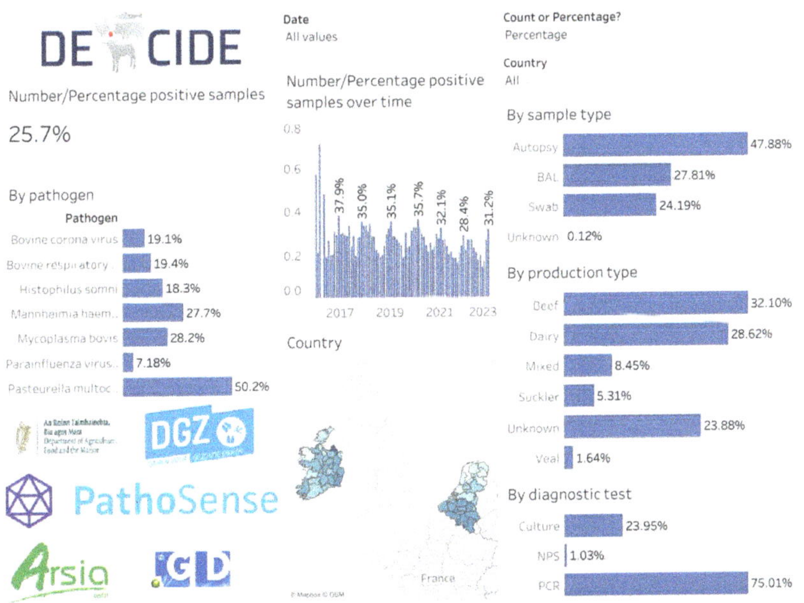

Figure 11 Cattle barometer use case, focusing on respiratory infectious diseases.

from the data related to respiratory infectious diseases in cattle. This tool is intended to be used by veterinary professionals, researchers, and livestock industry stakeholders such as farmers and producers keen on monitoring and mitigating the impact of diseases on cattle population, as shown in Fig. 11. The following are the different types of users who interact with a system or tool to achieve specific goals in future.

Veterinarians: Veterinary professionals can use the cattle barometer to monitor disease trends in specific regions, make informed decisions about vaccination strategies, and allocate resources to high-risk areas.

Researchers: Researchers can leverage the cattle barometer to identify patterns and contributing factors to disease outbreaks, facilitating the development of targeted interventions and prevention strategies.

Stakeholders: Stakeholders in the livestock industry, including farmers and producers, can access actionable insights to optimize herd management practices, minimize disease-related losses, and ensure sustainable production.

5 Conclusion and future trends

In this chapter, we have presented an ontology-driven, knowledge-based framework for farm animal health management systems, with a specific focus on a cattle barometer use case. We recognize the significant challenges faced in data acquisition, integration, and analysis within this domain and highlight the importance of agricultural semantics in addressing these challenges. By leveraging semantic web technologies, our framework aims to improve data management, enable informed decision-making, and ultimately enhance animal health outcomes.

One of the key contributions of our framework is the development of the LHO, which serves as a standardized framework for representing and organizing cattle health-related knowledge. LHO extends the existing DECIDE ontology and provides a species-specific ontology tailored specifically for cattle. This ontology-driven approach ensures standardization, integrity, interoperability, and reusability of data within farm animal health management systems. By making LHO available on Bioportal, we aim to promote its use and encourage community feedback in the field of farm animal health management.

To demonstrate the effectiveness of our ODKFADM, we have conducted a comprehensive case study focused on disease surveillance in cattle. The case study shows the practical application of our framework in improving disease monitoring, gathering feedback from users, and evaluating the impact of the framework on cattle health and management outcomes. The SPARQL query results highlight the potential of our approach in the field of farm animal health management.

In the future, we are extending the LHO to include species-specific ontologies for other farm animals such as salmon, pigs, and poultry. By doing so, we will further enhance the semantic interoperability and applicability of our framework across different livestock sectors. This expansion will enable a broader range of stakeholders to leverage the benefits of our ontology-driven approach, promoting data standardization and advanced analytical reasoning in the management of various farm animal health systems.

In conclusion, ontology-driven, knowledge-based framework for farm animal health management systems stands out from other animal livestock ontologies. Through practical use case applications and a farmer-friendly approach, it focuses on a cattle scenario, providing a standardized framework, species-specific LHO ontology, RDF data generation, and semantic interoperability. By adhering to FAIR guidelines and leveraging semantic web technologies, it enhances data accessibility, connectivity, and advanced analytical reasoning. Additionally, the framework prioritizes the protection and privacy of sensitive data in accordance with GDPR principles. The integration of data visualization tools facilitates data interpretation, empowering farmers and their veterinarians with valuable insights. Ultimately, this framework tackles data acquisition, integration, and analysis challenges, resulting in improved disease surveillance, early detection, and enhanced animal health outcomes.

However, we acknowledge that there are several challenges that need to be addressed in future research. These challenges include refining the ontology and improving its coverage, developing efficient data acquisition and integration mechanisms, enhancing analytical reasoning capabilities, and ensuring the scalability and sustainability of the framework. By tackling these challenges, we aim to continuously improve the effectiveness and practicality of our framework, advancing the field of farm animal health management and contributing to the overall well-being of livestock. Integration of federated learning techniques is another exciting future research within our framework. By incorporating federated learning, we can tap into the collective intelligence and data resources of multiple stakeholders while maintaining data privacy and security. This approach has the potential to further enhance disease surveillance, early detection, and decision-making in farm animal health management systems, ultimately leading to improved animal health outcomes in the agriculture.

6 Acknowledgements

The DECIDE project (https://decideproject.eu/) has received funding from the European Unions Horizon 2020 research and innovation program under grant agreement no. 101000494.

We would like to thank DGZ (https://www.dgz.be/mijn-dgz), The Royal GD (https://www.gddiergezondheid.nl/Diergezondheid/Onderzoek),Pathosen (https://www.pathosense.com/en/howpathosenseworks), ARSIA (https://www.arsia.be/), and Irish Laboratory (https://www.gov.ie/ga/eagraiocht/an-roinn-talmhaiochta-bia-agus-mara/) for their valuable data contributions.

7 References

Ali, W., Saleem, M., Yao, B., Hogan, A. and Ngomo, A.-C. N. 2022. A survey of RDF stores & SPARQL engines for querying knowledge graphs. *The VLDB Journal*, 31, 1-26.

Babcock, S., Beverley, J., Cowell, L. G. and Smith, B. 2021. The infectious disease ontology in the age of COVID-19. *Journal of Biomedical Semantics* 12(1), 1-20.

Bahlo, C., Dahlhaus, P., Thompson, H. and Trotter, M. 2019. The role of interoperable data standards in precision livestock farming in extensive livestock systems: a review. *Computers and Electronics in Agriculture* 156, 459-466.

Bhat, S. A. and Huang, N.-F. 2021. Big data and ai revolution in precision agriculture: survey and challenges. *IEEE Access* 9, 110209-110222.

Bullock, D. S., Boerngen, M., Tao, H., Maxwell, B., Luck, J. D., Shiratsuchi, L., Puntel, L. and Martin, N. F. 2019. The data-intensive farm management project: changing agronomic research through on-farm precision experimentation. *Agronomy Journal* 111(6), 2736-2746.

Chen, X., Jia, S. and Xiang, Y. 2020. A review: knowledge reasoning over knowledge graph. *Expert Systems with Applications* 141, 112948.

Cocca, G. and Lindberg, A. 2019. Overview of data analysis and reporting tools in the context of the SIGMA project. *EFSA Supporting Publications* 16(8), 1690E.

Cowell, L. G. and Smith, B. 2010. Infectious disease ontology,Infectious Disease Informatics. 373-395,Springer.https://doi.org/10.1007/978-1-4419-1327-2_19

Dórea, F. C., Vial, F., Hammar, K., Lindberg, A., Lambrix, P., Blomqvist, E. and Revie, C. W. 2019. Drivers for the development of an Animal Health Surveillance Ontology (AHSO). *Preventive Veterinary Medicine* 166, 39-48.

Fote, F. N., Roukh, A., Mahmoudi, S., Mahmoudi, S. A. and Debauche, O. 2020. Toward a big data knowledge-base management system for precision livestock farming. *Procedia Computer Science* 177, 136-142.

Gangadharan, V. and Gupta, D. 2020. Recognizing named entities in agriculture documents using LDA based topic modelling techniques. *Procedia Computer Science* 171, 1337-1345.

Golik, W., Dameron, O., Bugeon, J., Fatet, A., Hue, I., Hurtaud, C., Reichstadt, M., Salaün, M.-C., Vernet, J. and Joret, L. 2012. ATOL: the multi-species livestock trait ontology. In: Metadata and Semantics Research: 6th Research Conference, MTSR 2012, Cádiz, Spain, November 28-30, 2012. Proceedings 6. Springer, 289-300.

Gómez-Pérez, A. and Corcho, O. 2002. Ontology languages for the semantic web. *IEEE Intelligent Systems* 17(1), 54-60.

Hermans, K., Opsomer, G., Ranst, B. V. and Hostens, M. 2018. Promises and challenges of big data associated with automated dairy cow welfare assessment. In: *Animal Welfare in a Changing World*. CAB International Wallingford, A. Butterworth.

Heward-Mills, D. 2017. Visualizing SPARQL results in tableau. Available at: https://medium.com/virtuoso-blog/virtuoso-tableau-sparql-f9411852a87d (Accessed on 22 June 2023).

Hogan, A., Blomqvist, E., Cochez, M., D'amato, C., Melo, G. D., Gutierrez, C., Kirrane, S., Gayo, J. E. L., Navigli, R. and Neumaier, S. 2021. Knowledge graphs. *ACM Computing Surveys (CSUR)* 54, 1-37.

Hu, G., Do, D. N., Gray, J. and Miar, Y. 2020. Selection for favorable health traits: a potential approach to cope with diseases in farm animals. *Animals* 10(9), 1717.

Khatoon, P. S. and Ahmed, M. 2022. Importance of semantic interoperability in smart agriculture systems. *Transactions on Emerging Telecommunications Technologies* 33(5), e4448.

Kollia, I., Glimm, B. and Horrocks, I. 2011. SPARQL query answering over OWL ontologies. In: The Semantic Web: Research and Applications: 8th Extended Semantic Web Conference, ESWC 2011, Heraklion, Crete, Greece, May 29-June 2, 2011. Proceedings, Part I 8. Springer, 382-396.

Malhotra, A., Younesi, E., Gündel, M., Müller, B., Heneka, M. T. and Hofmann-Apitius, M. 2014. ADO: a disease ontology representing the domain knowledge specific to Alzheimer's disease. *Alzheimer's and Dementia* 10(2), 238-246.

Meyer, A., Faverjon, C., Hostens, M., Stegeman, A. and Cameron, A. 2021. Systematic review of the status of veterinary epidemiological research in two species regarding the FAIR guiding principles. *BMC Veterinary Research* 17(1), 270.

Monteiro, A., Santos, S. and Gonçalves, P. 2021. Precision agriculture for crop and livestock farming–brief review. *Animals* 11(8), 2345.

Murali, E. and Anouncia, S. M. 2022. An ontology-based knowledge mining model for effective exploitation of agro information. *IETE Journal of Research*, 68,1-18.

Newton, J. E., Nettle, R. and Pryce, J. E. 2020. Farming smarter with big data: insights from the case of Australia's national dairy herd milk recording scheme. *Agricultural Systems* 181, 102811.

Niloofar, P., Francis, D. P., Lazarova-Molnar, S., Vulpe, A., Vochin, M.-C., Suciu, G., Balanescu, M., Anestis, V. and Bartzanas, T. 2021. Data-driven decision support in livestock farming for improved animal health, welfare and greenhouse gas emissions: overview and challenges. *Computers and Electronics in Agriculture* 190, 106406.

Niloofar, P., Lazarova-Molnar, S., Francis, D. P., Vulpe, A., Suciu, G. and Balanescu, M. 2020. Modeling and simulation for decision support in precision livestock farming. *Winter Simulation Conference (WSC)*. IEEE, 2601-2612.

Peyre, M., Hoinville, L., Njoroge, J., Cameron, A., Traon, D., Goutard, F., Calba, C., Grosbois, V., Delabouglise, A., Varant, V., Drewe, J., Pfeiffer, D. and Häsler, B. 2019. The RISKSUR EVA Tool (Survtool): a tool for the integrated evaluation of animal health surveillance systems. *Preventive Veterinary Medicine* 173, 104777.

Rajbhandari, S. and Keizer, J. 2012. The AGROVOC concept scheme-a walkthrough. *Journal of Integrative Agriculture* 11(5), 694-699.

Ren, G., Lin, T., Ying, Y., Chowdhary, G. and Ting, K. C. 2020. Agricultural robotics research applicable to poultry production: a review. *Computers and Electronics in Agriculture* 169, 105216.

Rosa, G. J. 2021. *Grand Challenge in Precision Livestock Farming*. Frontiers in Animal Science. 2, 650324. doi: 10.3389/fanim.2021.650324

Roussaki, I., Doolin, K., Skarmeta, A., Routis, G., Lopez-Morales, J. A., Claffey, E., Mora, M. and Martinez, J. A. 2023. Building an interoperable space for smart agriculture. *Digital Communications and Networks* 9(1), 183–193.

Saha, S. 2021. *Towards a Formal Ontology to Support Knowledge Sharing across Product Design and Manufacturing*. Doctoral dissertation, Coventry University.

Sahoo, S. S., Halb, W., Hellmann, S., Idehen, K., Thibodeau Jr., T., Auer, S., Sequeda, J. and Ezzat, A. 2009. A survey of current approaches for mapping of relational databases to RDF. *W3C RDB2RDF Incubator Group Report* 1, 113–130.

Symeonaki, E., Arvanitis, K. G., Piromalis, D., Tseles, D. and Balafoutis, A. T. 2022. Ontology-based IoT middleware approach for smart livestock farming toward agriculture 4.0: a case study for controlling thermal environment in a pig facility. *Agronomy* 12(3), 750.

Van Schaik, G., Hostens, M., Faverjon, C., Jensen, D. B., Kristensen, A. R., Ezanno, P., Frössling, J., Dórea, F., Jensen, B.-B., Carmo, L. P., Steeneveld, W., Rushton, J., Gilbert, W., Bearth, A., Siegrist, M., Kaler, J., Ripperger, J., Siehler, J., de Wit, S., Garcia-Morante, B., Segalés, J., Pardon, B., Bokma, J. and Nielen, M. 2023. The DECIDE Project: from surveillance data to decision-support for farmers and veterinarians. *Open Research Europe* 3, 82.

Vasconcelos Gioia, G., Lamielle, G., Aguanno, R., Elmasry, I., Mouillé, B., De Battisti, C., Angot, A., Ewann, F., Sivignon, A. and Donachie, D. 2021. Informing resilience building: FAO's Surveillance Evaluation Tool (SET) biothreat detection module will help assess national capacities to detect agro-terrorism and agro-crime. *One Health Outlook* 3, 1–13.

Villeneuve, É., Abi Akle, A. A., Merlo, C., Masson, D., Terrasson, G. and Llaria, A. 2019. Decision support in precision sheep farming. *IFAC-PapersOnLine* 51(34), 236-241.

Chapter 5

Using data mining techniques for decision support in agriculture: support vector machines

Wu Caicong, China Agricultural University, China

1 Introduction

Data classification is a common task in machine learning. The support vector machine (SVM) is a powerful classification algorithm to achieve this goal. In machine learning, SVM is a supervised learning model with associated learning algorithms that analyzes data for classification and regression analysis. SVM is one of the most robust prediction methods in machine learning.

In SVM, data points are presented as p-dimensional vectors in order to find a $(p - 1)$ dimensional hyperplane (essentially a decision boundary) for linear classification and to determine the category into which to fit new data points. The central idea is to find a hyperplane with a maximum margin, which is the maximum distance between data points of different categories. This hyperplane is called the maximum margin hyperplane. The associated classifier is called the maximum margin classifier. SVM achieves this by constructing hyperplanes in high-dimensional or infinite-dimensional spaces, which are suitable for classification, regression, or detection of external values. The purpose is to

http://dx.doi.org/10.19103/AS.2023.0132.08

identify hyperplanes with the highest margin, as higher margins typically result in lower generalization errors.

Although the original problem can be expressed in a space of finite dimensions, it often turns out that the distinguishable set is not linearly separable in this space. For this reason, some have proposed mapping the original limited space to a higher-dimensional space, which can make separation in this space easier.

To keep the computational load reasonable, the mappings used by SVM schemes are designed to ensure that dot products of pairs of input data vectors may be computed easily in terms of the variables in the original space, by defining them in terms of a kernel function $k(x, y)$ selected to suit the problem. A hyperplane in high-dimensional space is defined as a set of points whose dot product with vectors in that space is constant, where such a vector set is the orthogonal (and therefore the minimum) vector set that defines the hyperplane.

The vectors defining the hyperplanes can be chosen to be linear combinations with parameters α_i of images of feature vectors x_i that occur in the data base. With this choice of a hyperplane, the points x in the feature space that are mapped into the hyperplane are defined by the relation $\sum_i \alpha_i k(x_i, x) = \text{constant}$. Note that if $k(x, y)$ becomes small as y grows further away from x, each term in the sum measures the degree of closeness of the test point x to the corresponding data base point x_i.

In this way, the sum of the above kernels can be used to measure the relative proximity of each test point to data points from one or another set to be distinguished. Please note that the set of points x mapped to any hyperplane can be very complex, allowing for more complex distinctions between completely non-convex sets in the original space.

2 The support vector machine model

The SVM algorithm was invented by Vladimir N. Vapnik and Alexey Y. Chervonenkis in 1964 and rapidly developed in the 1990s, resulting in a series of improved and extended algorithms. SVM is a classifier developed from the generalized portrait algorithm in pattern recognition. In 1992, Bernhard Boser, Isabelle Guyon, and Vladimir Vapnik suggested a way to create noninear classifiers by applying the kernel trick to maximum-margin hyperplanes (Boser et al., 1992). The 'soft margin' incarnation, as is commonly used in software packages, was proposed by Corinna Cortes and Vapnik in 1993 and published in 1995 (Cortes and Vapnik, 1995).

Given a set of learning instances, each labeled as belonging to one of two categories, the SVM learning algorithm constructs a model and assigns the new instance to one or another category, making it a non-probabilistic binary linear classifier (although some methods, such as Platt scaling, use SVM in probabilistic classification parameters). SVM maps training examples to a point in space to maximize the spatial width between two categories. Then map the new examples to the same space and predict which side of the gap they belong to base on which category they belong to.

In addition to performing linear classification, SVMs can efficiently perform a nonlinear classification using what is called the kernel trick, implicitly mapping their inputs into high-dimensional feature spaces.

The support vector clustering algorithm created by Hava Siegelmann and Vladimir Vapnik applies vector support statistical data developed in SVM algorithms to classify unlabeled data. These datasets require unsupervised learning methods, attempting to find natural data clusters and then mapping new data based on these clusters.

2.1 Hard-margin support vector machines

If the training data is linearly separable, we can select two parallel hyperplanes that separate the two classes of data, so that the distance between them is as large as possible. The region bounded by these two hyperplanes is called the 'margin', and the maximum-margin hyperplane is the hyperplane that lies halfway between them.

Given input data and learning objectives in classification problems: $X = \{X_1, \dots, X_N\}$, $y = \{y_1, \dots, y_N\}$. Each sample of input data contains multiple features, which form a feature space: $X_i = \{x_1, \dots, x_n\} \in \chi$, and the learning objective is binary variables $X_i = \{-1, \dots, 1\}$ represents negative class and positive class. With a normalized or standardized dataset, these hyperplanes can be described by the equations:

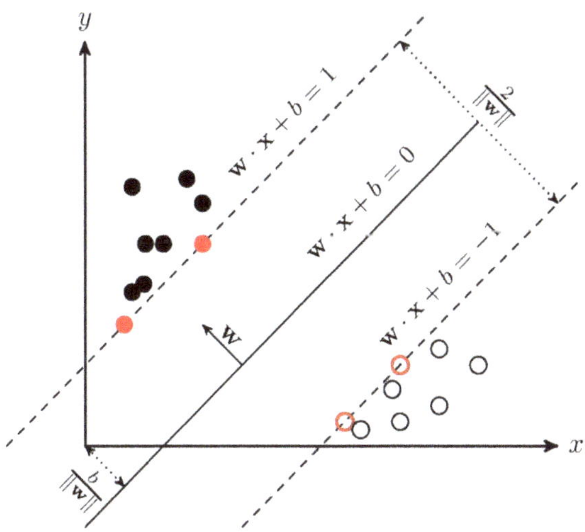

$\mathbf{w}^T\mathbf{x}+b=1$ (anything on or above this boundary is of one class, with label 1)

and

$\mathbf{w}^T\mathbf{x}+b=-1$ (anything on or below this boundary is of the other class, with label -1).

Geometrically, the distance between these two hyperplanes is $\frac{2}{\|\mathbf{w}\|}$, so to maximize the distance between the planes, we want to minimize $\|\mathbf{w}\|$. The distance is computed using the distance from a point to a plane equation. We also have to prevent data points from falling into the margin, we add the following constraint: for each i either

$\mathbf{w}^T\mathbf{x}_i+b\geq 1$, if $y_i=1$,

Or

$\mathbf{w}^T\mathbf{x}_i+b\leq -1$, if $y_i=-1$.

These constraints state that each data point must lie on the correct side of the margin.

This can be rewritten as

$y_i\left(\mathbf{w}^T\mathbf{x}_i+b\right)\geq 1$, for all $1\leq i\leq n$.

We can put this together to get the optimization problem

$$\underset{\mathbf{w},b}{\text{minimize}}\ \frac{1}{2}\mathbf{w}^2$$
$$\text{subject to}\ y_i\left(\mathbf{w}^T\mathbf{x}_i+b\right)\geq 1\,\forall i\in\{1,\ldots,n\}$$

$$\underset{\mathbf{w},b}{\text{maximize}}\ \frac{2}{\mathbf{w}}$$
$$\text{subject to}\ y_i\left(\mathbf{w}^T\mathbf{x}_i+b\right)\geq 1\,\forall i\in\{1,\ldots,n\}$$

The $\mathbf{w}$ and b that solve this problem determine our classifier, $\mathbf{x}\mapsto \text{sgn}\left(\mathbf{w}^T\mathbf{x}+b\right)$ where $\text{sgn}(\cdot)$ is the sign function. An important consequence of this geometric description is that the max-margin hyperplane is completely determined by those $\mathbf{x}_i$ that lie nearest to it. These $\mathbf{x}_i$ are called support vectors:

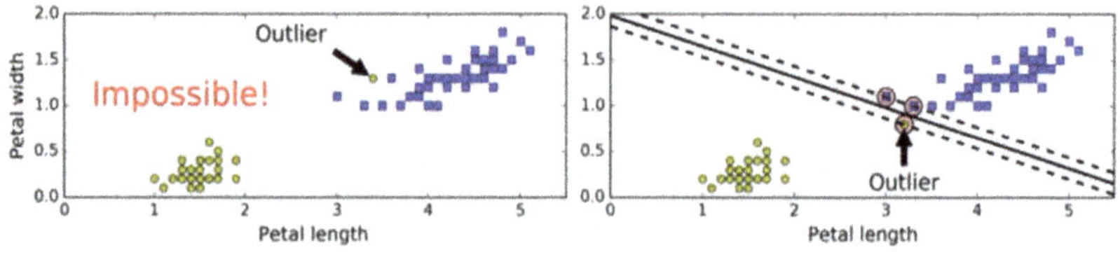

2.2 Soft-margin support vector machines

To extend SVM to cases in which the data are not linearly separable, the hinge loss function is helpful:

$$\max(0,1-y^i+(w^T x_i+b))$$

Note that y^i is the *i*th target (i.e. in this case, 1 or −1), an is the *i*th output. This function is zero if the constraint in (1) is satisfied, in other words, if $\mathbf{x}_i$ lies on the correct side of the margin. For data on the wrong side of the margin, the function's value is proportional to the distance from the margin. The goal of the optimization then is to minimize.

$$\lambda \| w \|^2+[1/n \sum^n_{(i=1)} \max(0,1-y_i(w^T x_i+b))]$$

where the parameter $\lambda>0$ determines the trade-off between increasing the margin size and ensuring that the $\mathbf{x}_i$ lie on the correct side of the margin. By deconstructing the hinge loss, this optimization problem can be optimized into the following:

$$\underset{\mathbf{w},b,\zeta}{\text{minimize}} \ \frac{1}{2}|\mathbf{w}|^2+C\sum_{i=1}^{n}\zeta_i$$

$$\textit{subject to}\ y_i\left(\mathbf{w}^T\mathbf{x}_i+b\right)\geq 1-\zeta_i, \zeta_i\geq 0 \forall i\in\{1,\ldots,n\}$$

Thus, for large values of C, it will behave similar to the hard-margin SVM, if the input data are linearly classifiable, but we will still learn if a classification rule is viable or not. (λ is inversely related to C, e.g. in LIBSVM.)

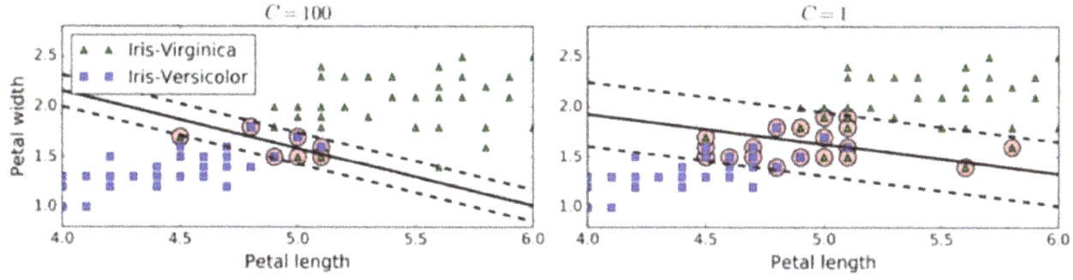

2.3 Primal and dual support vector machines

Minimizing (2) can be rewritten as a constrained optimization problem with a differentiable objective function in the following way:

For each $i\in\{1,\ldots,n\}$, we introduce a variable $\zeta_i=max\left(0,1-y_i\left(\mathbf{w}^T\mathbf{x}_i+b\right)\right)$. Note that ζ_i is the smallest nonnegative number satisfying $y_i\left(\mathbf{w}^T\mathbf{x}_i+b\right)\geq 1-\zeta_i$.

Thus, we can rewrite the optimization problem as follows:

$$\text{minimize} \frac{1}{n}\sum_{i=1}^{n}\zeta_i + \lambda \| \mathbf{w} \|^2$$

subject to $y_i\left(\mathbf{w}^T\mathbf{x}_i + b\right) \geq 1 - \zeta_i$ and $\zeta_i \geq 0$, for all i.

This is called the primal problem.

By solving for the Lagrangian dual of the above problem, one obtains the simplified problem:

$$\text{maximize} f(c_1 \ldots c_n) = \sum_{i=1}^{n} c_i - \frac{1}{2}\sum_{i=1}^{n}\sum_{j=1}^{n}[y_i c_i \left(\mathbf{x}_i^T \mathbf{x}_j\right) y_j c_j],$$

$$\text{subject to} \sum_{i=1}^{n} c_i y_i = 0, \text{and } 0 \leq c_i \leq \frac{1}{2n\lambda} \text{ for all } i.$$

This is called the dual problem. Since the dual maximization problem is a quadratic function of the c_i subject to linear constraints, it is efficiently solvable by quadratic programming algorithms.

Here, the variables c_i are defined such that:

$$\mathbf{w} = \sum_{i=1}^{n} c_i y_i \mathbf{x}_i .$$

Moreover, $c_i = 0$ exactly when $\mathbf{x}_i$ lies on the correct side of the margin, and $0 < c_i < (2n\lambda)^{-1}$ when $\mathbf{x}_i$ lies on the margin's boundary. It follows that $\mathbf{w}$ can be written as a linear combination of the support vectors.

The offset, b, can be recovered by finding an $\mathbf{x}_i$ on the margin's boundary and solving

$$y_i\left(\mathbf{w}^T\mathbf{x}_i + b\right) = 1 \Leftrightarrow b = \mathbf{w}^T\mathbf{x}_i - y_i .$$

(Note that $y_i^{-1} = y_i$ since $y_i = \pm 1$.)

2.4 The kernel trick in support vector machines

We are given a training dataset of n points of the form $(\mathbf{x}_1, y_1), \ldots, (\mathbf{x}_n, y_n)$, where the y_i are either 1 or −1, each indicating the class to which the point $\mathbf{x}_i$ belongs. Each $\mathbf{x}_i$ is a p-dimensional real vector. We want to find the 'maximum-margin hyperplane' that divides the group of points $\mathbf{x}_i$ for which $y_i = 1$ from the group of points for which $y_i = -1$, which is defined so that the distance between the hyperplane and the nearest point $\mathbf{x}_i$ from either group is maximized. Any hyperplane can be written as the set of points $\mathbf{x}$ satisfying:

$$\mathbf{w}^T\mathbf{x} + b = 0,$$

where $\mathbf{w}$ is the (not necessarily normalized) normal vector to the hyperplane. This is much like Hesse normal form, except that $\mathbf{w}$ is not necessarily a unit vector. The parameter $\frac{b}{\|\mathbf{w}\|}$ determines the offset of the hyperplane from the origin along the normal vector $\mathbf{w}$.

The maximum margin hyperplane algorithm proposed by Vapnik in 1963 constructed a linear classifier. However, in 1992, Bernhard Boser, Isabelle Guyon, and Vladimir Vapnik proposed a method that applies kernel techniques (initially proposed by Aizerman et al.) to the maximum margin hyperplane to create nonlinear classifiers. This allows the algorithm to adjust the maximum margin hyperplane in the transformed feature space. Transformations can be nonlinear and the transformation space has large dimensions; Although the classifier is hyperplane in the transformed feature space, it may be nonlinear in the original input space.

It is noteworthy that working in a higher-dimensional feature space increases the generalization error of SVM, although given enough samples the algorithm still performs well.

Some common kernels include:

- Polynomial (homogeneous): $k(\mathbf{x}_i,\mathbf{x}_j)=(\mathbf{x}_i\cdot\mathbf{x}_j)^d$. Particularly, when $d=1$, this becomes the linear kernel.
- Polynomial (inhomogeneous): $k(\mathbf{x}_i,\mathbf{x}_j)=(\mathbf{x}_i\cdot\mathbf{x}_j+r)^d$.
- Gaussian radial basis function: $k(\mathbf{x}_i,\mathbf{x}_j)=\exp(-\gamma\mathbf{x}_i-\mathbf{x}_j{}^2)$ for $\gamma>0$. Sometimes parametrized using $\gamma=\frac{1}{2\sigma^2}$
- Sigmoid function (Hyperbolic tangent): $k(\mathbf{x}_\mathrm{i},\mathbf{x}_\mathrm{j})=\tanh(\kappa\mathbf{x}_i\cdot\mathbf{x}_j+c)$ for some (not every) $\kappa>0$ and $c<0$.

The kernel is related to the transform $\varphi(\mathbf{x}_i)$ by the equation $k(\mathbf{x}_i,\mathbf{x}_j)=\varphi(\mathbf{x}_i)\cdot\varphi(\mathbf{x}_\mathrm{j})$. The value $\mathbf{w}$ is also in the transformed space, with $\mathbf{w}=\sum_i\alpha_i y_i\varphi(\mathbf{x}_i)$. Dot products with $\mathbf{w}$ for classification can again be computed by the kernel trick, i.e. $\mathbf{w}\cdot\varphi(\mathbf{x})=\sum_i\alpha_i y_i k(\mathbf{x}_i,\mathbf{x})$.

Suppose now that we would like to learn a nonlinear classification rule which corresponds to a linear classification rule for the transformed data points $\varphi(\mathbf{x}_i)$. Moreover, we are given a kernel function k which satisfies:

$$k(\mathbf{x}_i,\mathbf{x}_j)=\varphi(\mathbf{x}_i)\cdot\varphi(\mathbf{x}_j).$$

We know the classification vector $\mathbf{w}$ in the transformed space satisfies:

$$\mathbf{w}=\sum_{i=1}^{n}c_i y_i\varphi(\mathbf{x}_i),$$

where the c_i are obtained by solving the optimization problem:

$$\text{maximize} f(c_1 \ldots c_n) = \sum_{i=1}^{n} c_i - \frac{1}{2}\sum_{i=1}^{n}\sum_{j=1}^{n}[y_i c_i (\varphi(\mathbf{x}_i)\cdot\varphi(\mathbf{x}_j)) y_j c_j]$$

$$= \sum_{i=1}^{n} c_i - \frac{1}{2}\sum_{i=1}^{n}\sum_{j=1}^{n}[y_i c_i k(\mathbf{x}_i, \mathbf{x}_j) y_j c_j]$$

$$\text{subject to} \sum_{i=1}^{n} c_i y_i = 0, \text{ and } 0 \le c_i \le \frac{1}{2n\lambda} \text{ for all } i.$$

The coefficients c_i can be solved for using quadratic programming, as before. Again, we can find some index i such that $0 < c_i < (2n\lambda)^{-1}$, so that $\varphi(\mathbf{x}_i)$ lies on the boundary of the margin in the transformed space, and then solve:

$$b = \mathbf{w}^T\varphi(\mathbf{x}_i) - y_i = \left[\sum_{j=1}^{n} c_j y_j \varphi(\mathbf{x}_j)\cdot\varphi(\mathbf{x}_i)\right] - y_i$$

$$= \left[\sum_{j=1}^{n} c_j y_j k(\mathbf{x}_j, \mathbf{x}_i)\right] - y_i$$

Finally,

$$\mathbf{z} \mapsto \operatorname{sgn}(\mathbf{w}^T\varphi(\mathbf{z}) - b) = \operatorname{sgn}\left(\left[\sum_{i=1}^{n} c_i y_i k(\mathbf{x}_i, \mathbf{z})\right] - b\right).$$

3 Developments in support vector machines algorithms

The recent algorithms used to find SVM classifiers include sub-gradient descent and coordinate descent. It has been proven that both techniques have significant advantages over traditional methods when dealing with large dispersed datasets sub-gradient methods are particularly effective when there are many training examples, and coordinate descent methods are particularly effective when the dimensionality of the feature space is high.

3.1 Sub-gradient descent

Sub-gradient descent algorithms for the SVM work directly with the expression

$$f(\mathbf{w}, b) = \left[\frac{1}{n}\sum_{i=1}^{n} max\left(0, 1 - y_i\left(\mathbf{w}^T\mathbf{x}_i + b\right)\right)\right] + \lambda \| \mathbf{w} \|^2 .$$

Note that f is a convex function of $\mathbf{w}$ and b. As such, traditional gradient descent (or SGD) methods can be adapted, where instead of taking a step in the direction of the function's gradient, a step is taken in the direction of a vector selected from the function's sub-gradient. This approach has the advantage that, for certain implementations, the number of iterations does not scale with n, the number of data points (Shalev-Shwartz et al., 2011).

Coordinate descent algorithms for the SVM work from the dual problem:

$$\text{maximize } f(c_1 \ldots c_n) = \sum_{i=1}^{n} c_i - \frac{1}{2} \sum_{i=1}^{n} \sum_{j=1}^{n} [y_i c_i (x_i \cdot x_j) y_j c_j],$$

$$\text{subject to } \sum_{i=1}^{n} c_i y_i = 0, \text{ and } 0 \leq c_i \leq \frac{1}{2n\lambda} \text{ for all } i.$$

For each $i \in \{1, \ldots, n\}$, iteratively, the coefficient c_i is adjusted in the direction of $\partial f / \partial c_i$. Then, the resulting vector of coefficients $(c_1', \ldots, c_n')$ is projected onto the nearest vector of coefficients that satisfies the given constraints. (Typically, Euclidean distances are used.) The process is then repeated until a near-optimal vector of coefficients is obtained. The resulting algorithm is extremely fast in practice, although few performance guarantees have been proven.(Hsieh et al., 2008).

The above soft margin SVM is an example of the empirical risk minimization (ERM) algorithm for hinge loss. From this perspective, SVM belong to a natural class of statistical inference algorithms, many of which have unique features due to hinge loss behavior. This viewpoint helps us to better understand the working mode and reasons of SVM, and enables us to better analyze its statistical characteristics.

3.2 Risk minimization

In supervised learning, one is given a set of training examples $X_1 \ldots X_n$ with labels $y_1 \ldots y_n$, and wishes to predict y_{n+1} given X_{n+1}. To do so one forms a hypothesis, f, such that $f(X_{n+1})$ is a 'good' approximation of y_{n+1}. A 'good' approximation is usually defined with the help of a loss function, $\ell(y, z)$, which characterizes how bad z is as a prediction of y. We would then like to choose a hypothesis that minimizes the expected risk:

$$\varepsilon(f) = \mathbb{E}\left[\ell\left(y_{n+1}, f(X_{n+1})\right)\right].$$

In most cases, we do not know the joint distribution of X_{n+1}, y_{n+1} outright. In these cases, a common strategy is to choose the hypothesis that minimizes the empirical risk

$$\hat{\varepsilon}(f) = \frac{1}{n}\sum_{k=1}^{n}\ell(y_k, f(X_k)).$$

Under certain assumptions about the sequence of random variables X_k, y_k (e.g. that they are generated by a finite Markov process), if the set of hypotheses being considered is small enough, the minimizer of the empirical risk will closely approximate the minimizer of the expected risk as n grows large. This approach is called ERM.

3.3 Regularization and stability

In order for the minimization problem to have a well-defined solution, we have to place constraints on the set H of hypotheses being considered. If H is a normed space (as is the case for SVM), a particularly effective technique is to consider only those hypotheses f for which $\| f \| < k$. This is equivalent to imposing a regularization penalty $R(f) = \lambda_k \| f \|$, and solving the new optimization problem:

$$\hat{f} = \arg\min_{f \in H} \hat{\varepsilon}(f) + R(f).$$

This approach is called Tikhonov regularization. More generally, $R(f)$ can be some measure of the complexity of the hypothesis f, so that simpler hypotheses are preferred.

3.4 SVM and the hinge loss

Recall that the (soft-margin) SVM classifier $\hat{\mathbf{w}}, b : \mathbf{x} \mapsto \operatorname{sgn}(\hat{\mathbf{w}}^T\mathbf{x} - b)$ is chosen to minimize the following expression:

$$\left[\frac{1}{n}\sum_{i=1}^{n} max\left(0, 1 - y_i\left(\mathbf{w}^T\mathbf{x} + b\right)\right)\right] + \lambda \| \mathbf{w} \|^2.$$

In light of the above discussion, we see that the SVM technique is equivalent to ERM with Tikhonov regularization, where in this case the loss function is the hinge loss:

$$\ell(y, z) = max(0, 1 - yz).$$

From this perspective, SVM is closely related to other fundamental classification algorithms such as regularized least-squares and logistic regression. The difference between the three lies in the choice of loss function: regularized least-squares amounts to ERM with the square-loss, $\ell_{sq}(y, z) = (y - z)^2$; logistic regression employs the log-loss:

$$\ell_{\log}(y, z) = \ln\left(1 + e^{-yz}\right).$$

3.5 Target functions

The difference between the hinge loss and these other loss functions is best stated in terms of target function - the function that minimizes expected risk for a given pair of random variables X, y.

In particular, let y_x denote y conditional on the event that $X = x$. In the classification setting, we have

$$y_x = \begin{cases} 1 \; \textit{with probability } p_x \\ -1 \; \textit{with probability } 1\text{-}p_x \end{cases}$$

The optimal classifier is therefore,

$$f^*(x) = \begin{cases} 1 \; \textit{if } p_x \geq \frac{1}{2} \\ -1 \; \textit{otherwise} \end{cases}$$

For the square loss, the target function is the conditional expectation function, $f_{sq}(x) = \mathbb{E}[y_x]$; For the logistic loss, it is the logit function, $f_{log}(x) = \ln(p_x / (1-p_x))$. While both of these target functions yield the correct classifier, as $\text{sgn}(f_{sq}) = \text{sgn}(f_{log}) = f^*$, they give us more information than we need. In fact, they give us enough information to completely describe the distribution of y_x.

3.6 Extensions: multiclass support vector machines

Multiclass SVM aims to assign labels to instances by using SVM, where the labels are drawn from a finite set of several elements. The dominant approach for doing so is to reduce the single multiclass problem into multiple binary classification problems (Duan and Keerthi, 2005).

One approach is to build a binary classifier to distinguish one label from other labels (one-on-one) or each pair of classes (one-on-one). In a one-on-one situation, the classification of new instances is carried out through a win-win strategy, where the classifier with the highest output function assigns the category (it is important to calibrate the output function to produce comparable scores). For the one-on-one method, classification is achieved through a maximum winning vote strategy, where each classifier assigns an instance to one of two classes, then adds one vote to the assigned class's vote, and finally, the class with the most votes determine the classification of the instance.

3.7 Extensions: Bayesian support vector machines

In 2011, Polson and Scott demonstrated that SVM achieved Bayesian interpretation through data extension technology. In this method, SVM is considered as a graphical model (where parameters are connected through probability distributions). This extended view allows for the application of Bayesian techniques to SVM, such as flexible feature modeling, automatic adjustment of hyperparameters, and quantification of prediction uncertainty. Recently, Florian Wenzel developed an extensible version of Bayes SVM, which allows for the application of Bayes SVM to big data. Florian Wenzel has developed two different versions, the Variational Inference (VI) scheme of Bayesian kernel SVM and the Random Version (SVI) of Linear Bayesian SVM.

4 Challenges in using support vector machines

The SVM is an important algorithm in the field of machine learning and is widely used for its excellent performance and solid mathematical foundation. It has achieved success in many fields by constructing optimal hyperplanes in high-dimensional spaces for data classification. However, SVM also has some drawbacks, such as long training time, sensitivity to missing values, and multi-class challenges. In this section, we will delve into these shortcomings of SVM. The following sections introduce other models and technologies to compensate for them.

4.1 Long training time

SVM needs to calculate the distance between each pair of data points during training, and its computational complexity is usually O(n^2) or O(n^3), where n is the number of training samples. In high-dimensional spaces, this computational burden becomes even heavier, so the training time of SVM increases significantly when dealing with large-scale data sets, potentially leading to unacceptable latency. To deal with this problem, researchers have proposed some methods, such as kernel approximation techniques and random sampling subsets to speed up the training process. In addition, the distributed SVM framework can also be used to parallelize training, thereby reducing training time.

4.2 Sensitivity to missing values

SVM usually requires that all features have valid values, and data integrity between features needs to be ensured. However, in real-world data, missing values are common, and SVM is very sensitive to missing values. The methods for handling missing values include deleting missing value samples, replacing missing values with filled values (such as mean or median), or using specialized missing value processing techniques. Choosing an appropriate method depends on the nature of the data and the distribution of missing values.

4.3 Difficulties in handling large-scale multi-class problems

The original SVM was designed for binary classification problems. To use SVM on multi-category problems, it is usually necessary to adopt strategies such as one-vs-rest or one-vs-one to combine multiple categories. Convert this problem into several sub-problems of binary classification. This will increase the complexity of the model and may require a significant amount of training time and storage space. In addition, class imbalance issues require additional processing to ensure that the model has appropriate performance between classes. To address these issues, researchers have proposed multi-class SVM extension and class balancing methods to simplify the processing of multi-class classification problems.

4.4 Adjusting parameters is complicated

SVM has some key parameters, including the regularization parameter C, the type and parameters of the kernel function, etc. The choice of these parameters has an important impact on the performance of the model, but adjusting these parameters is not easy. The optimal combination of search parameters usually requires techniques such as cross-validation, which may require significant resources and computational time. Similarly, the selection of kernel functions is also an important issue, as different kernel functions can be applied to different types of data. Therefore, appropriate parameter adjustment and kernel function selection need to be carried out by combining domain knowledge and experiments.

5 Solutions to support vector machine shortcomings: other models

To compensate for some of the shortcomings of the SVM model, researchers and data scientists have developed a range of other models that may be more suitable in specific situations and can solve the problems that SVM has. Here are some solution-based models and how they address the shortcomings of SVM.

5.1 Random forest

Random forest, a widely employed ensemble learning technique based on decision trees, offers a host of distinctive advantages. It constitutes a collection of independent decision trees, which are subsequently integrated through methods like voting or averaging. This integration fortifies the model's stability and its capacity to generalize from the data. What sets random forest apart is its infusion of randomness in the training process. This involves the random selection of training samples and features, a strategy that effectively mitigates the risk of overfitting. Each decision tree is trained on a distinct subset, diversifying the model's learning.

Additionally, random forests have the remarkable ability to estimate the importance of each feature in a classification or regression task, providing valuable insights into the underlying data and model dynamics. Notably, random forests address several limitations associated with SVM. They demonstrate efficiency, resilience in handling missing values, and a natural aptitude for multi-category classification tasks. Particularly on extensive datasets, they can construct multiple decision trees in parallel, significantly reducing training time. Moreover, they can seamlessly accommodate missing data without the need for additional preprocessing steps. This ensemble method excels in multi-category classification, as each decision tree is proficient in managing multiple categories, simplifying the overall process.

5.2 Gradient Boosting Trees

Gradient Boosting Trees emerge as a powerful ensemble learning technique, distinguished by several key strengths. They operate by constructing multiple decision trees and combining them for enhanced model performance. Each tree in this ensemble is trained to address the errors of its predecessor, resulting in a gradual refinement of the model. This iterative optimization process equips gradient-boosted trees to excel in handling complex problems. Moreover, akin to random forests, they demonstrate an innate ability to seamlessly handle missing values in the data without necessitating additional preprocessing steps, both during training and prediction phases. Notably, when compared to SVM, gradient-boosted trees prove highly efficient in optimizing iteratively, making them particularly adept at addressing intricate nonlinear challenges and large-scale datasets. They often strike a superior balance between training time and performance.

Additionally, they exhibit insensitivity to missing values and naturally support multi-category classification tasks, eliminating the need for cumbersome strategies. These attributes have propelled gradient-boosted trees to widespread popularity in the realms of data science competitions and real-world applications, as they offer an efficient alternative that effectively addresses various limitations associated with SVMs.

5.3 Deep learning models

Deep learning models, encompassing architectures like convolutional neural networks (CNN) for image and spatial data, as well as recurrent neural networks (RNN) for sequential data such as text and time series, excel in tackling intricate problems and processing extensive datasets. CNNs leverage multiple convolutional and pooling layers to adeptly capture intricate features and patterns within images. On the other hand, RNNs, equipped with looping structures, exhibit a unique ability to factor in contextual information in sequences.

Notably, deep learning models address the limitations of SVM effectively. They have extraordinary feature learning abilities, allowing them to autonomously extract features from raw data without the need for laborious engineering work. It has been proven that this skill is very valuable for dealing with complex nonlinear challenges and large-scale datasets, and can reduce the long-term training of SVM. CNN and RNN models have inherent nonlinear modeling capabilities, gradually extracting and fusing features through multi-layer neural networks to better adapt to nonlinear data. Deep learning is suitable for large datasets, combined with its ability to recognize complex models, overcoming the performance limitations of SVM in the field of big data. Overall, deep learning is an excellent tool that outperforms SVM in all aspects and is a powerful solution to complex problems.

5.4 Conclusion

Although SVM is a powerful machine learning algorithm known for its robustness and effectiveness in various scenarios, it does come with some limitations. These include its tendency for long training times, sensitivity to missing values, and potential challenges when dealing with large-scale multi-class problems. However, it is important to acknowledge the strengths of SVM, particularly its ability to find optimal hyperplanes in high-dimensional spaces, making it highly effective for tasks like binary classification and support for nonlinear classification through the use of kernels.

To address these limitations and explore alternative approaches, one can consider employing other models like random forests, gradient boosting trees, deep learning models, and XGBoost. The choice of model should be guided by the specific nature and requirements of the problem at hand, as well as the characteristics of the available datasets and computational resources. In practical applications, it is crucial to carefully evaluate and select the most appropriate model based on the specific situation. While SVM remains a powerful choice in many situations, there are instances where other models may offer more suitability and effectiveness for the given task.

6 Case study: identifying damage from cotton spider mites

Cotton spiders or leaf mites are one of the most challenging pests in cotton cultivation. They pose significant difficulties in control due to their small size, ability to conceal themselves within cotton plants, wide dissemination, and long pest life cycle (Yang Lili, 2019). Traditional methods of identifying the level of cotton leaf mite damage rely heavily on experienced agricultural professionals conducting on-site manual inspections. However, this approach has limitations such as cost, time, limited detection coverage, and the inability to accurately assess the extent of damage across entire cotton fields. This, in turn, can adversely affect cotton yield and quality.

The timely and accurate recognition of the level of cotton leaf mite damage is crucial for precise monitoring and management of pest infestations in cotton fields. In this study, under natural conditions, images of cotton leaves with varying levels of cotton leaf mite damage were captured using smartphones. Image segmentation algorithms were applied to separate the cotton leaves from the background. Subsequently, color, texture, and boundary features were extracted from the segmented cotton leaves. Comparative analysis was

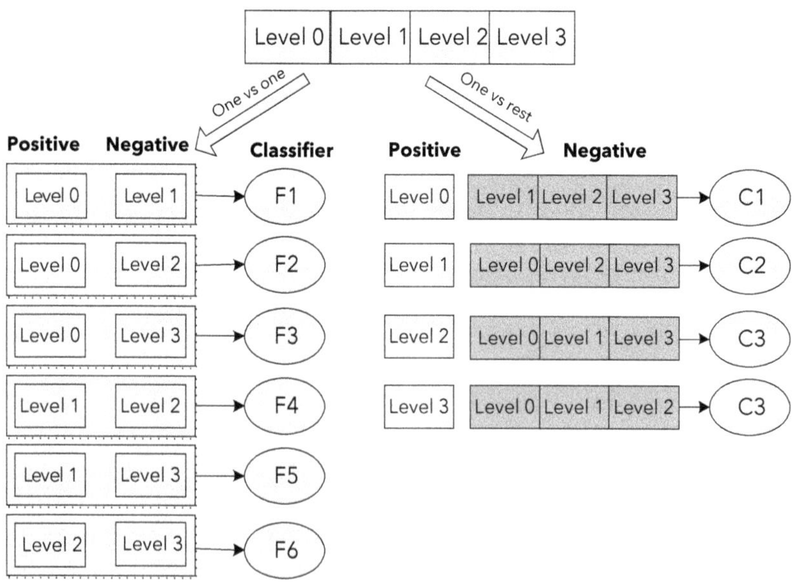

Figure 1 Multi-classification split strategy.

performed to identify a distinct set of features that can differentiate between different levels of cotton leaf mite damage.

This study employs a combination of the SVM and AdaBoost algorithms to train a classification model for identifying the level of cotton leaf mite damage. Initially, the SVM algorithm is chosen for training the classification model. This study adopts a four-class split strategy, specifically 'one-vs-one' and 'one-vs-rest', where levels 0 through 3 represent the levels 0 through 3. The generated classifiers' outcomes corresponding to this strategy are illustrated in Fig. 1.

The 'one-vs-rest' method, when training a classifier, requires the use of all training samples. Compared to the 'one-vs-one' approach, it involves more training data, which results in greater time and space overhead. Additionally, constructing training data for positive and negative classes in the 'one-vs-rest' method can lead to imbalanced data, adversely affecting the classifier's generalization ability. In this study, experiments were conducted using the 'one-vs-rest' method. To address the data imbalance issue, the smote algorithm was employed to synthetically augment the minority class data.

The final SVM algorithm achieved an average recognition accuracy of 70% across the four damage levels of cotton leaf mite damage, which was considered unsatisfactory. As a result, this study adopted the 'one-vs-one' split strategy, generating six classifiers for the four damage levels of cotton leaf mite damage. These classifiers were then combined using a voting method to determine the damage level of cotton leaf mite damage, resulting in improved recognition results.

This study utilized the SVM algorithm to train and generate three classifiers, namely, F2, F3, and F6. Additionally, the AdaBoost algorithm was employed to train and create three classifiers, namely, F1, F4, and F5. These six classifiers were then integrated, and the final classification result was obtained using a voting method, as shown in Table 1. From the table, it is evident that the ensemble classifier produced by combining SVM and AdaBoost achieved an average recognition accuracy of 88.75%.

In terms of cotton leaf mite damage level recognition, the algorithm presented in this chapter achieved an average recognition accuracy of 88.75% with a processing time of 0.389 s. Compared to classifiers generated by SVM,

Table 1 Cotton spider mites damage level recognition result generated by combination classifier

Damage level	Number of samples	Level 0	Level 1	Level 2	Level 3	Accuracy/%
0	20	18	1	1	0	90
1	20	2	17	1	0	85
2	20	0	1	17	2	85
3	20	1	0	0	9	95

the recognition accuracy improved by 12.5 percentage points. In comparison to classifiers generated by AdaBoost, there was a reduction in processing time by 0.092 s. Compared to classifiers generated by the BP neural network, improvements were observed in both recognition accuracy and processing time.

7 Case study: identifying turning trajectories of a wheat harvester

Equipping agricultural machinery with GNSS terminals enables real-time tracking, with trajectory data revealing machinery behavior and field path distribution – crucial for operation scheduling and analysis (Yang Lili, 2023). Identifying turning trajectories is essential for improving efficiency and planning routes, as studies indicate that turning consumes approximately 40% of field operation time, with a U-turn being particularly efficient. Recognizing turning trajectories allows for efficiency calculations, route planning, and field shape assessments. This case study proposes a machine learning-based algorithm for recognizing turning the trajectories of wheat harvesters in the field.

Initially, a two-step K-means clustering and a three-step correction are employed to identify X-turn, abnormal working, and working trajectory. To further classify U-turn trajectories from the working trajectory, a U-turn recognition algorithm based on a SVM model is constructed, with three subsequent

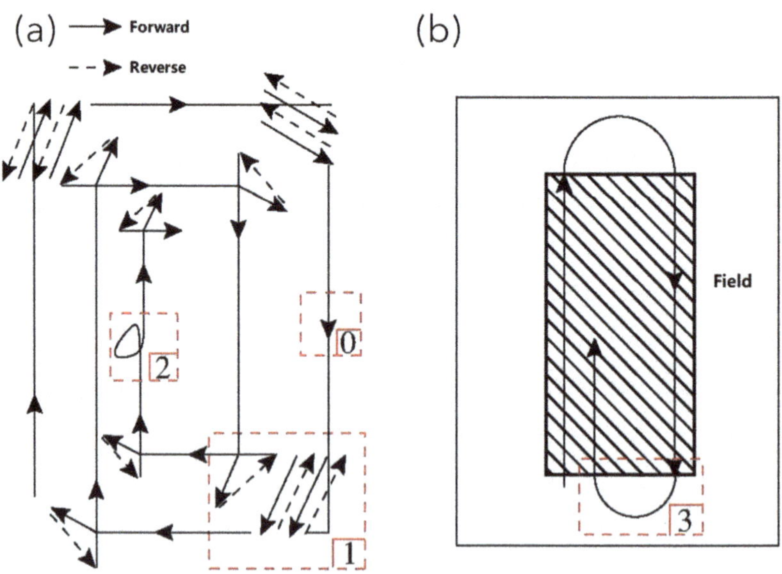

Figure 2 Schematic diagram of harvesting path.

correction steps applied to the preliminary identification results. Ultimately, the algorithm successfully identifies X-turn, abnormal working, U-turn, and working trajectory within the wheat harvester's field, achieving a 94% F1 score. The F1 score for data with intervals ranging from 1 to 5 s is above 90%, enabling detailed field trajectory segmentation. Utilizing the effective operation trajectory obtained by removing turning and abnormal trajectories, field area calculations through distance algorithms result in a 12.76% reduction in calculation errors compared to using raw trajectories. This research provides valuable insights for precision management based on extensive agricultural machinery trajectories.

In wheat harvesting, there are two common turning methods: X-turn and U-turn, illustrated in Fig. 2a. X-turn involves multiple forward-reverse-forward movements by agricultural machinery during field operations, as seen in dashed-line box 1. Abnormal working happens when the machinery deviates from its path, creates detours or irregular behaviors, and then returns to the original path, as shown in dashed-line box 2. U-turn directly changes direction without reversing, represented by dashed-line box 3 in Fig. 2b. All other trajectories, excluding turning trajectories and abnormal working, are considered working trajectories, denoted by dashed-line box 0. This chapter focuses on recognizing these four trajectories in wheat harvesting.

The X-turn recognition algorithm consists of two modules: clustering and correction. The clustering module conducts two K-means clustering rounds. The first uses angle features, dividing the trajectory into turns and other points. The second round uses distance features to classify working and X-turn trajectories. The correction module refines results with M1, M2, and M3. M1 calculates angle statistics for each segment to identify cluster centroids, measuring distances to correct misclassifications. M2 adjusts abnormal operation segments mistakenly identified as X-turns if the direction change is under 10°. M3 defines X-turn start and endpoints, categorizing nearby points within 3.4 m, defining X-turn boundaries. This ensures precise X-turn identification.

Table 2 Trajectory recognition algorithm results (%)

Algorithm	P	R	F1 score
D- K-means	53	62	55
D-K-means+M1	53	63	56
D-K-means+M1+M2	85	94	88
D-K-means+M1+M2+M3	93	96	95
SVM	78	86	81
SVM+S1	80	94	84
SVM+S1+S2	90	95	92
SVM+S1+S2+S3	92	95	94

U-turn recognition uses SVM and three corrections: S1, S2, and S3. S1 refines U-turn boundaries using circumscribed circle curvature. S2 adjusts U-turn identification based on the time between U-turns and the nearest preceding X-turn. S3 corrects U-turn recognition by analyzing angle changes before and after the turn. These corrections enhance U-turn trajectory recognition alongside SVM in agricultural machinery operations Table 2.

The experimental results, as shown in Table 2, indicate a significant improvement in the initial recognition and subsequent correction of X-turn trajectories, with the F1 value increasing from 55% to 95%. In the case of U-turn trajectory recognition, the SVM-based recognition resulted in an accuracy of 81%, which was further improved to 94% after correction. This demonstrates that the X-turn trajectory recognition algorithm and the U-turn trajectory recognition algorithm are highly effective in identifying the four types of trajectories in wheat harvester field operations.

8 Case study: tractor emission prediction

Emissions from agricultural machinery should not be underestimated. According to statistics from the Chinese Ministry of Ecology and Environment, in 2022, NOx emissions from agricultural machinery in China exceeded 149,000 tons, and estimated CO_2 emissions reached 166 million tons based on fuel consumption. China officially implemented the 'National IV' emissions

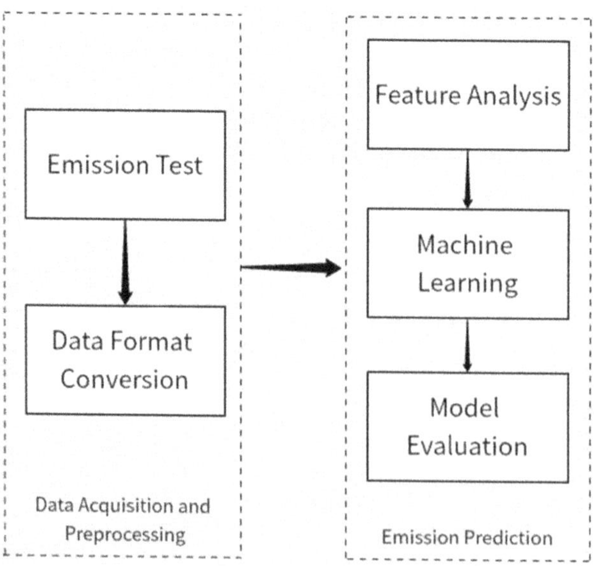

Figure 3 Technology roadmap of this case.

standard on December 1, 2022, imposing stricter requirements on pollutant emissions from agricultural machinery.

Reducing emissions from agricultural machinery is a key focus of China's efforts to combat air pollution, and the introduction of the 'peak carbon dioxide emissions and carbon-neutral' goals underscores China's determination to reduce carbon emissions and address greenhouse gas effects. Through emission forecasting, we can provide insights for the formulation of emission reduction standards and measures, better achieve energy efficiency and emission reduction in agricultural machinery, offer data references for the analysis of air pollutants and greenhouse gas emissions, and contribute to the sustainable development of agriculture. In this case, a portable emission measurement system (PEMS) was used to conduct emission tests on tractors, and an emission forecasting model was developed based on tractor engine parameters and environmental factors.

The technical roadmap for this case, as shown in Fig. 3, is divided into two main parts:

- Data acquisition and preprocessing; and
- Emission prediction.

Emission tests for tractors were conducted using PEMS. Four tractors were included in the experiment, with power ratings of 51.5 kW, 58.8 kW, 88.2 kW, and 177 kW. To obtain a dataset for training the prediction model, the engine speed and instantaneous fuel consumption of the tractors were controlled by adjusting the traction force applied to the tractors. This adjustment allowed for changes in the tractor's emission rate, providing ample training samples for the model. PEMS continuously monitored the tractor's instantaneous emissions.

The HORIBA OBS-ONE equipment used in the experiment consisted of modules including gas analyzers, GPS, tailpipe adapters, and more. It accurately monitored the exhaust emissions of the tractors, including CO_2 and NOx, as well as key engine parameters such as engine speed, torque, and instantaneous fuel consumption. Environmental parameters like temperature and humidity, which directly affect the fuel combustion process and energy efficiency, were also measured. These parameters served as features and inputs for the model.

The SVM model was trained using the available data. SVM can learn the relationship between these features and emission rates, enabling accurate predictions for unknown data. In this case, different feature combinations were explored to determine which features had the best effect on agricultural machinery emission prediction, thus enhancing the model's prediction accuracy. Evaluation metrics for the prediction model included R^2 (coefficient of determination) and MAPE (mean absolute percentage error).

Based on experimentation, the combination of environmental temperature, humidity, engine speed, torque, and instantaneous fuel consumption yielded the best results for predicting tractor emissions. The results showed that the SVM algorithm achieved an R^2 value of 0.949 and an MAPE of 0.060. As shown in Table 2, compared to other machine learning algorithms such as ridge regression, random forest, and LightGBM, the SVM algorithm achieved superior performance, making it more accurate in completing the task of agricultural machinery emission prediction.

9 Case study: spring wheat yield prediction

Wheat is the world's most widely cultivated crop. Wheat breeding aims for adaptable, disease-resistant, high-yield, and stable germplasm (Shafiee, 2021). Traditional pedigree breeding takes about 10–12 years before new lines become breeding parents. To address the slow and resource-intensive nature of traditional phenotyping, interest has grown in using unmanned aerial vehicles (UAV) for imaging-based phenotyping. UAV imagery in visible and near-infrared wavelengths (Vis-NIR) provides various spectral indices for estimating vegetation properties like chlorophyll levels, pigments, and leaf area index (LAI). UAV images have shown promise in predicting grain yield, a critical trait in plant phenotyping and precision agriculture for crops like wheat, maize, and rice.

This study explores two machine learning-based regression methods for wheat grain yield prediction using vegetation indices extracted from UAV images (Shafiee et al., 2021). The study assesses the effectiveness of support vector regression (SVR) combined with sequential forward selection (SFS) and compares results with the LASSO regressor featuring an internal feature selector. The models were evaluated using grain yield data from 600 spring wheat plots planted in South-Eastern Norway in 2018. Five spectral bands and three vegetation indices (NDVI, EVI, and MTCI) were extracted from multispectral images at three different stages between heading and maturity. These features served as input data for the SVR model, with model hyperparameters optimized through grid search.

This study aimed to:

1 Use sequential forward selection for variable selection for yield prediction and compare it with the LASSO variable selector;
2 Use grid search for tuning the hypermeters of the model and selecting the better machine learning regression model for yield prediction; and
3 Compare the explanatory power of three major VIs for grain yield prediction.

Table 3 Selected indices by LASSO and their importance scores, displayed as absolute regression coefficients with grain yield.

Date	Selected indices	Importance
26.06.2018 (47 days after sowing)	NDVI	0.54
	MTCI	0.5
	EVI	0.14
02.07.2018 (54 days after sowing)	NDVI	0.71
	MTCI	0.42
	EVI	0.008
19.07.2018 (70 days after sowing)	NDVI	1
	MTCI	0.11

LASSO was first formulated by Robert Tibshirani in 1996. It is a powerful method that performs two main tasks including regularization and feature selection. The LASSO method puts a constraint on some of the absolute values of the model parameters. The sum must be less than a fixed value (upper bound). In order to do so, the method applies a shrinkage (or regularization) process where it penalizes the coefficient of regression variables, thereby shrinking some of them to zero.

During the feature selection process, the variables that still have non-zero coefficients after the shrinkage process are selected to be part of the model. The goal of the process is to minimize the prediction error. In this study, the tuning parameter for regularization amount control was determined using 10-fold cross-validation. The length of the path (min-lambda/max-lambda) was chosen to be 0.001 and 100 default values along the regularization pass were tested to find the best lambda value Table 3.

SVR is an application of SVM for regression cases. Depending on whether the relationship between data is linear or nonlinear, the SVR is using a linear or nonlinear kernel function. In this study, grid search was used to determine the appropriate kernel function and the *C* parameter for the SVR model, and the gamma parameter was set on default value. Two different types of kernels including radial basis function (RBF) kernel and linear kernel along with *C* values of 1, 50, 100, 200, 300, 400, 500, 600, 700, and 1000 were tested.

Sequential feature selection trims a starting *d*-dimensional feature space to a smaller *k*-dimensional subspace (where $k < d$). It aims to automatically choose the most relevant features, enhancing computational efficiency and minimizing model errors by eliminating irrelevant features or noise. Two approaches exist: forward (SFS) and backward (SBS). SFS starts with one feature, models the data, and selects the one with the highest accuracy or performance metric. This repeats for a user-set number of features. SBS begins with all features and removes the one with the least impact on performance, again down to a set number of

Table 4 LASSO and SVR regression results for the training set using selected vegetation indices by LASSO and SFS, respectively

Date	Regression model					
	SVR (selected features by SFS)			LASSO (selected features by LASSO)		
	R^2	MSE	Training time (s)	R^2	MSE	Training time (s)
26.06.2018 (47 days after sowing)	0.86	0.21	730	0.82	0.27	<2
02.07.2018 (54 days after sowing)	0.84	0.23	750	0.82	0.29	<2
19.07.2018 (70 days after sowing)	0.89	0.17	800	0.86	0.21	<2
All dates together (Pool)	0.90	0.16	1100	0.90	0.15	<3

features. To ensure full coverage, you can add or remove previously selected features, known as sequential forward floating selection (SFFS) or sequential backward floating selection (SBFS). The ML methods' performance is assessed using mean square error (MSE) and coefficient of determination (R^2) Table 4.

The results indicate that machine learning has a great ability to predict wheat grain yield using spectral indices from images. There are some advantages to some of those methods over others. In this study, we have shown that SVR in combination with SFS is a robust method for grain yield prediction based on UAV imagery and that LASSO regression yields similar results with much less computation time for feature selection. The explanatory capability of vegetation indices for grain yield prediction differs during the season. Some indices are more predictive of grain yield in the early growth stages and some later in the season. Multi-temporal remote sensing data provides more accurate prediction than single-temporal data. Comprehensive research is still needed in different environments to obtain firm conclusions on the most prominent vegetation indices for grain yield prediction.

10 Conclusion and future trends

In summary, SVM is a powerful supervised learning model primarily used for classification and regression analysis. The core idea of SVM is to construct a maximum-margin hyperplane for classification. When the training data is linearly separable, SVM selects two parallel hyperplanes to separate data points of different classes while maximizing the distance between the two hyperplanes; this region is referred to as the 'margin.' The maximum-margin hyperplane lies in the middle of the two parallel hyperplanes and is considered the optimal decision boundary for classification.

The optimization goal of SVM is to minimize the norm of the weight vector while ensuring that each training sample is on the correct side of the hyperplane through a set of constraints. These constraints require that each sample point satisfies an inequality to ensure correct classification. If the data is not perfectly linearly separable, SVM introduces slack variables that allow some points to fall within the margin with a penalty. This is known as soft-margin SVM, where a regularization parameter balances margin size and classification errors.

SVM not only handles linear classification but also performs nonlinear classification through kernel tricks. Kernel tricks allow SVM to map data into a high-dimensional feature space, effectively transforming nonlinear problems into linearly separable ones. Commonly used kernel functions include polynomial kernels, Gaussian radial basis functions, and hyperbolic tangent functions, among others.

Furthermore, SVM has multiple extensions and variations, including support vector clustering, multi-class SVM, transudative SVM, structured SVM, and Bayesian SVM. These extensions enhance the applicability of SVM and enable its use in various tasks such as unsupervised learning, semi-supervised learning, structured prediction, and big data analytics. However, SVM has some inherent limitations, including the requirement for fully labeled data and less interpretable model parameters. Therefore, when applying SVM, careful selection of the kernel function, parameter tuning, consideration of data characteristics, and task requirements are necessary.

This chapter introduces several examples of SVM algorithms applied in the agricultural domain, including the use of SVM for cotton leaf mite damage level identification, farm machinery operation trajectory classification, and agricultural machinery emissions prediction tasks. Through feature extraction and data analysis, SVM algorithms have yielded satisfactory results in identifying crop pests, tractor turns trajectories, and agricultural machinery emissions, thereby improving agricultural production efficiency, resource utilization efficiency, and environmental protection. These applications demonstrate the significant potential of SVM algorithms in addressing complex issues and promoting sustainable agricultural development.

With the continuous development of SVM, the adoption of SVM, a powerful machine learning algorithm, has increased significantly in various fields. Known for their versatility and effectiveness, SVMs also have great potential to drive future advancements in agricultural research. As a powerful tool in the agricultural field, SVM demonstrates a wide range of application possibilities.

In precision agriculture, weed identification for variable spraying is crucial. Tang Jinglei and others propose an SVM-based method using decision trees on visible/near-infrared images. This approach combines spectral characteristics and a thresholding technique for segmentation. By integrating features and a maximum voting mechanism, classification accuracy is enhanced, meeting real-time demands. Suresh et al. introduce the closed-loop irrigation decision support system (CLISS) for water-efficient irrigation. With global water scarcity, this system is essential. Traditional irrigation consumes 70% of global water. CLISS uses autonomous sensors for data collection and employs SVM, PLSR, and ANFIS for precise irrigation decisions. Validation in semi-arid areas confirms the model's ability to predict crop water needs and schedule irrigation, offering detailed production forecasts. Extensive research affirms SVM's pivotal role in yield prediction, disease identification, and decision support systems in precision agriculture. It significantly improves production efficiency and resource utilization.

As industries grow, listed companies play a crucial role. With their expansion, financial risks become a significant concern. This is particularly pronounced in agriculture due to its inherent uncertainty and seasonality. Therefore, establishing a suitable financial early warning model is imperative for predicting, analyzing, and managing financial risks. Some scholars have employed SVM and other techniques to create early warning models for agricultural listed companies. They utilized the SVM algorithm, principal component analysis, and the *Z*-value method to assess the financial risk status of 48 agricultural listed companies in 2020. By employing these methods, they identified 28 key financial indicators and achieved a high prediction rate of 97.9%. The main identified cause of financial risks was the accumulation of risks and inadequate risk resistance.

Soil erosion, the displacement of soil particles, poses significant threats like flooding, desertification, and harm to infrastructure, impacting agriculture. Alireza et al. employed four models - maximum entropy, artificial neural network, SVM, and general linear model - to map gully erosion susceptibility in the Golestan Dam watershed. This robust model aids in soil and water conservation.

In conclusion, sustainable agricultural development requires judicious resource use, technological advancements, and institutional reforms to meet present and future demands. SVM models play a role in achieving these goals.

11 Where to look for further information

Further reading

- Bennett, K. P. and Campbell, C. (2000). Support vector machines: Hype or hallelujah? (PDF). *SIGKDD Explorations* 2(2), 1–13. doi:10.1145/380995.380999. S2CID 207753020.

- Cristianini, N. and Shawe-Taylor, J. (2000). *An Introduction to Support Vector Machines and Other Kernel-Based Learning Methods*. Cambridge University Press. ISBN 0-521-78019-5.
- Fradkin, D. and Muchnik, I. (2006). Support vector machines for classification (PDF). In Abello, J. and Carmode, G. (Eds.), *Discrete Methods in Epidemiology*. DIMACS Series in Discrete Mathematics and Theoretical Computer Science. Vol. 70, pp. 13–20.
- Joachims, T. (1998). Text categorization with support vector machines: Learning with many relevant features. In Nédellec, C. and Rouveirol, C. (Eds.), *Machine Learning: ECML-98*. Lecture Notes in Computer Science. Vol. 1398. Berlin: Springer. pp. 137–142. doi:10.1007/BFb0026683. ISBN 978-3-540-64417-0. S2CID 2427083.
- Ivanciuc, O. (2007). Applications of support vector machines in chemistry (PDF). *Reviews in Computational Chemistry* 23, 291–400. doi:10.1002/9780470116449.ch6. ISBN 9780470116449.
- James, G., Witten, D., Hastie, T. and Tibshirani, R. (2013). Support vector machines (PDF). In *An Introduction to Statistical Learning: with Applications in R*. New York: Springer, pp. 337–372. ISBN 978-1-4614-7137-0.
- Schölkopf, B. and Smola, A. J. (2002). *Learning with Kernels*. Cambridge, MA: MIT Press. ISBN 0-262-19475-9.
- Steinwart, I. and Christmann, A. (2008). *Support Vector Machines*. New York: Springer. ISBN 978-0-387-77241-7.
- Theodoridis, S. and Koutroumbas, K. (2009). *Pattern Recognition* (4th ed.). Academic Press. ISBN 978-1-59749-272-0.

Libraries

- libsvm, LIBSVM is a popular library of SVM learners.
- liblinear is a library for large linear classification including some SVMs.
- SVM light is a collection of software tools for learning and classification using SVM.
- SVMJS live demo is a GUI demo for JavaScript implementation of SVMs.

12 References

Boser, B. E., Guyon, I. M. and Vapnik, V. N. 1992. A training algorithm for optimal margin classifiers. In: Proceedings of the Fifth Annual Workshop on Computational Learning Theory, COLT '92. Association for Computing Machinery, New York, NY, pp. 144–152. https://doi.org/10.1145/130385.130401.

Cortes, C. and Vapnik, V. 1995. Support-vector networks. *Machine Learning* 20(3), 273–297. https://doi.org/10.1007/BF00994018.

Duan, K.-B. and Keerthi, S. S. 2005. Which is the best multiclass SVM method? An empirical study. In: Multiple Classifier Systems. Presented at the International Workshop on

Multiple Classifier Systems, Springer, Berlin, pp. 278–285. https://doi.org/10.1007/11494683_28.

Hsieh, C.-J., Chang, K.-W., Lin, C.-J., Keerthi, S. S. and Sundararajan, S. 2008. A dual coordinate descent method for large-scale linear SVM. In: Proceedings of the 25th International Conference on Machine Learning, ICML '08. Association for Computing Machinery, New York, NY, pp. 408–415. https://doi.org/10.1145/1390156.1390208.

Shafiee, S., Lied, L. M. B., Burud, I., Dieseth, J. A., Alsheikh, M. K. and Lillemo, M. 2021. Sequential forward selection and support vector regression in comparison to LASSO regression for spring wheat yield prediction based on UAV imagery. 183. https://doi.org/10.1016/j.compag.2021.106036.

Shalev-Shwartz, S., Singer, Y., Srebro, N. and Cotter, A. 2011. Pegasos: primal estimated sub-gradient solver for SVM. *Mathematical Programming* 127(1), 3–30. https://doi.org/10.1007/s10107-010-0420-4.

Yang, L., Wang, X., Li, Y. and Wu, C. 2023 Identifying turning trajectories of wheat harvester based on machine learning. Transactions of the Chinese Society for Agricultural Machinery, 1-8.

Yang Lili, Z. D. 2019. Automatic recognition for cotton spider mites damage level based on SVM and AdaBoost. Nongye Jixie Xuebao/Transactions of the Chinese Society of Agricultural Machinery, 50.

Part 2

Case studies

Chapter 6

Developing decision support systems for irrigation/water management on farms

Fedro S. Zazueta, University of Florida, USA

1 Introduction

A decision support system (DSS) is a tool used by a decision-maker to make informed decisions and improve decision outcomes. A DSS can be presented in many forms, from a simple fact sheet to a computer-based system that uses communication technologies, data, documents, knowledge, and modeling to define problems and prescribe actions (Power and Heavin, 2017), including the use of automation and intelligent systems. A DSS can be applied at various stages of a decision-making process to describe a system, predict its behavior, conduct a diagnosis, and prescribe actions that change the system and its behavior in desired ways.

Computer-based DSSs have been used in agriculture for over 55 years. By the early 1980s, over 1500 computer programs were available for agriculture, forestry, food and nutrition, livestock, marketing, land investment analysis, and other areas (Strain, 1982). Examples of early DSSs in agriculture include spreadsheets, dedicated software applications, expert systems, and modeling and simulation. These have evolved with technology from simple stand-alone applications running on a computer to complex data-driven applications interacting on real time with other systems in the farm. Innovations such as mobile computing/communications, social media, agricultural drones, Internet of Things (IoT), big data, and cloud computing presented new challenges and

http://dx.doi.org/10.19103/AS.2023.0132.12

Table 1 Examples of technologies associated to the 4th industrial revolution (4IR)

Additive layer manufacturing (3D print)
Artificial intelligence and machine learning (AI/ML)
Autonomous vehicles
Big data
Blockchain
Brain computer interfaces (BCI)
Cloud and edge computing
Cloud computing
Cyberphysical systems
High-speed networks (e.g. 5G)
Internet of systems (IoS)
Internet of Things (IoT)
New biology
Quantum computing
Robots and cobots
Supercomputing
Virtual reality (VR) and augmented reality
Other

opportunities for agriculture to redefine and rethink the role of ICT toward achieving better farming outcomes (Xin and Zazueta, 2016). Currently, DSSs are being developed by researchers and practitioners based on technologies that create a promising future (Table 1). Blurring of the boundaries between these technologies will result in convergence of technologies that would bring about deep innovation (Lee, 2018).

A successful DSS is one that provides value to a management decision and is adopted by the stakeholder. Considerations in the design, development, and implementation and adoption of DSS have been discussed by several researchers (Zhai et al., 2020; Rose et al., 2016; Köksal and Tekinerdogan, 2019).

2 Developing a decision support system

There are a few general principles that must be adhered to when developing a DSS. The application must be:

1 Correct. A DSS must resolve the problem at hand or provide useful information conforming to fact from a trustworthy source.
2 Usable. A DSS is focused on improving outcomes and reducing management overhead. Thus, the design must be intuitive from the user perspective, easy to use, and allowing effortless understanding of the architecture and navigation throughout the application.
3 Convenient. Access to the application must be available at the point of use. The most general case is that the application be available anywhere, anytime, on any device, and produce a response in a reasonable

time. This is especially important for the design of applications that require access to data in the cloud or remote servers where there is no connectivity.

4 Efficient. The user should be able to accomplish the related tasks to the DSS and contribute to the decision on hand quickly and efficiently.

To achieve the above, a clear definition of the problem addressed by the DSS must be articulated, a known solution to the problem must be available, the bounds of its applicability must be clearly articulated, and the target audience's competency regarding the knowledge domain of the application must be known. For example, a common decision that must be made in the design of a drip system for fruit trees is the number of emitters per tree. Thus, the definition of the problem is reduced to the question, 'How many emitters per tree?' There are several approaches to determine the number of emitters, one of which is to wet a fraction of the root volume of the crop irrigated. If this approach was to be used, several methods can be used to determine the number of emitters. One solution is to use a dynamic equation to simulate the movement of water in the soil and determine what fraction of the root zone is wetted by a single emitter and calculate how many emitters are required to meet the stated constraint. If such a solution was used in a DSS, it must be clear to the user that the field conditions where this solution is to be applied must meet the assumptions made in developing the dynamic model.

Knowledge of the competence of the user for which the DSS is intended is essential to a successful DSS. It is also important to ensure user involvement during requirement analysis and testing phases of the project.

For example, if the question is, 'How much water should an irrigation system deliver?' A DSS to deliver water for residential turf in an urban setting is substantively different from that of a large citrus grove or golf course, where the tolerance related to quality and the economies of pumping water has a direct impact on the solvency of the business. Irrigation for a home turf may be addressed by giving recommendations on how to set an irrigation timer, while in a large agricultural production system, careful irrigation management must be carried out concurrently with the application of various chemicals through the irrigation system and other agronomic and environmental practices to ensure a product that is commercially viable. In this example, the level of competence required by the user (a typical homeowner vs. a grove manager) to set an irrigation timer is substantially different from that of developing a fertigation calendar.

Also, useful DSS software is the result of a development process subject to a life cycle. There are several models that address the software development life cycle (SDLC) (Saravanan et al., 2020). The selected model for managing

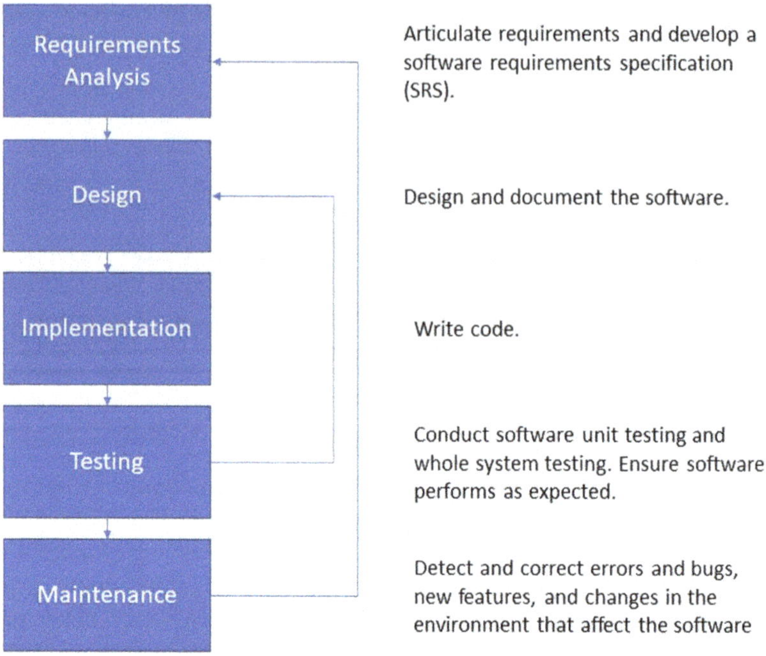

Figure 1 Engineering techniques are applied to produce quality-maintainable software. These software engineering techniques have evolved along the lines of a software development life cycle. Different approaches to the SDLC vary in their detailed approach but fall roughly into the general steps shown in this figure.

the SDLC of a DSS should be that which best fulfills the client's requirements (Kushwaha and Misra, 2006). Figure 1 shows the major phases in the SDLC. It is important for the development team to have skills related to software process modeling, life cycle, management, and software improvement and to recognize that this is an ongoing process (Kneuper, 2018). It is also important to follow standardized approaches (Information Technology Task Force, 2017) that ensure good life cycle practice, and focus on a product that addresses customer needs and is maintainable.

3 Water in the farm

Developing a DSS for farm water management requires detailed knowledge of the role of water in agricultural production. A farm is part of the hydrologic cycle, as a result, on-farm water management practices have an impact beyond the farm boundaries. Broadly, water in the farm can be classified as follows (Fig. 2):

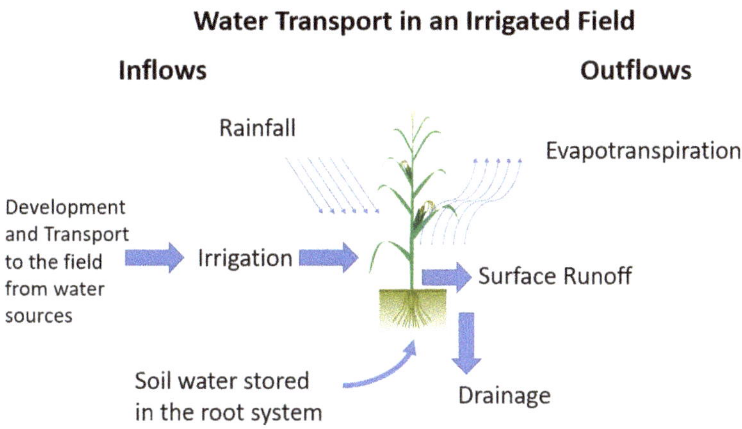

Figure 2 An irrigated field is a subsystem of the hydrologic cycle. Water may be obtained from surface or groundwater sources and conducted to the irrigated field where it is applied to the crop to supplement rainfall. Also, water may leave the system as evapotranspiration, runoff, or drainage. The primary role of irrigation is to maintain soil water within the root system to ensure healthy crop growth. Modern irrigation systems are also used to ensure proper plant nutrition and health through the application of dissolved chemicals.

3.1 Water sources

Water may be available to the farm from surface or underground sources. Surface water sources may include lakes, streams, or water conveyed to the farm by hydraulic infrastructure such as dams and canals, or overland pipes. Groundwater sources tap into underlying aquifers. To use and manage water in the farm, the temporal availability of water from these sources is an important factor in decisions related to the design of systems to control and manage water in the farm. Irrigation solutions at any given location will be affected in many ways by the availability of water on demand or by prescribed water delivery cycles imposed by existing infrastructure or regulations.

It is not only important to know the amount of water available over time, but also to know the biological and chemical properties of the water. Biological agents may include pathogens or other organisms that affect the performance of the system, especially when using recycled water (Hong and Moorman, 2005). Also, naturally occurring chemicals in the water may require treatment prior to use in irrigation to address problems related to corrosion of equipment (Rodrigues et al., 2020), effects on plant response (Valdez-Aguilar and Reed, 2007), clogging of the irrigation system (Haman, 2022), and interactions with fertilizers and other substances applied through the irrigation system (Bozkurt and Ozekici, 2006).

As competition for water among different sectors has increased, approaches to water resource management are changing (Gleik, 2009, 2018). Water withdrawals to supply domestic growth are stimulated by income growth

(Alcamo et al., 2010), requiring that water use in agriculture to become more efficient, especially in areas affected by climate change.

3.2 Inflows: irrigation and rainfall

Irrigation aims to supply the water deficit, required for healthy plant growth, resulting from the uneven distribution or insufficient quantity of rainfall. When crops are irrigated, the highest consumer of water on the farm is the irrigation process. Irrigation systems may apply water either on the surface or subsurface of the soil. The most common irrigation systems are variations of water applied at the surface of the soil that rely on gravity to distribute water across the field, water conducted in pressurized pipes that distribute water to large sections of the field (surface or subsurface), and systems that rely on managing a water table to supply water to the crop by capillarity. Concepts related to irrigation systems are described below.

3.3 Outflows: evapotranspiration, surface runoff, and drainage

After an irrigation event, water is lost in the field by surface runoff or drainage below the root zone (seepage). Thereafter, evapotranspiration takes place until the next irrigation or rainfall event, i.e. water is taken up by the plant from the soil and evaporated from the soil surface.

Excess water in the field must be removed to maintain a suitable crop growth environment (Evans and Fausey, 1999) by means of suitable drainage. Drainage problems can be broadly classified as surface or subsurface drainage. Excess irrigation water as well as the rainfall fraction that does not infiltrate into the soil must be removed. Also, if no natural drainage occurs, a path is required to transport excess water in the root zone and to maintain a salt balance (Letey et al., 2011). In addition, field conditions may require removal of water from sources such as seepage from adjoining bodies of water such as lakes or canals. Some components of the drainage system may vary from year to year depending on climate conditions. For example, precipitation abnormalities due to El Niño Southern Oscillation (ENSO) impact warm, dry, and rainy patterns in some parts of the world. In humid locations where this occurs, attention must be given to surface drainage to mitigate fungal and bacterial diseases, among other negative impacts on the crop (Knowx and Griffin, 2022).

3.4 Other uses

Other significant uses of large volumes of water include environmental control (Perry, 1998), such as the need for protection from freezes as shown in Fig. 3, or a mist system to maintain high humidity in a propagation greenhouse. Water

Figure 3 Overhead irrigation used to protect blueberries from freezing. Because the freezing point of a plant is lower than the freezing point of water, by applying sufficient water to maintain the temperature of water at the triple point (0°C) or above, plants survive a short freeze. Photo credit: Josh Wickman, IFAS, University of Florida.

on the farm has many uses other than irrigation, such as using it as a solvent for many types of chemical products, washing equipment, etc.

4 Irrigation systems

The goal of an irrigation system is to manage water in an agricultural production system that optimizes the value, quantity, and quality of crop production. For a given crop and growth environment, an irrigation system is used to:

1. Provide water to meet evapotranspiration demands of the crop;
2. Maintain an equilibrium of salts in the system;
3. Transport nutrients and other chemicals into the system; and
4. Maintain a controlled environment, such as air temperature and humidity.

Successful implementation of irrigation technologies requires the appropriate selection of irrigation system type for a given set of crop production conditions, design, installation, and management of the system, including maintenance and operation. The purpose of selection of system type, design, and installation is to ensure sufficient and timely water delivery to the cropped area in the most economical form possible, with the least environmental impact. Although tools for design of different irrigation system types have been available for many years (Zazueta and Smajstrla, 1989; Hutchinson et al., 1993; Strelkoff et al.,

2000; Brajović et al., 2015), DSSs focused primarily on the operation of irrigation systems, primarily irrigation scheduling. Management recommendations and related DSSs are made with the underlying assumption that a given irrigation system can deliver water uniformly on demand. However, in some hydro agricultural projects, water is delivered to farms on a rotating basis. A discussion of all irrigation system types is beyond the scope of this chapter.

5 Irrigation scheduling

From a water demand perspective, a fundamental problem of irrigation system management related to operation, not considering nutrient transport or other uses, is to determine (1) the amount of water that is to be delivered to the field and (2) when water is to be delivered to the field. In the past, several strategies were used successfully to address these two items and include (Fig. 4):

1. Water budgeting techniques based on mass conservation and soil dynamics (Heermann et al., 1990);
2. Soil moisture measurement (Nellingham, 2009; Yu et al., 2021);
3. Plant water stress indicators (Jones, 2004); and
4. Heuristic and other approaches (Naadimuthu et al., 1999).

Irrigated soils are a complex system that has physical, chemical, and biological components. As a result of this complexity assumptions are made when describing and modeling the system, which results in an incomplete description of the system and its governing equations. This is further complicated by the fact that in the field, boundary conditions are not simple, and most often, soil systems are not homogeneous and isotropic. Also, soil properties may change over time, depending on the interaction of the different components of the system. As a result, it is common to use more than one of the strategies above and described below. A challenge for the DSS developer is to hide this complexity behind a usable interface.

5.1 Water budgets based on mass conservation

Water budget methods are an application of the principle of mass conservation or a dynamic equation in an isothermal, homogeneous, and isotropic system. The control volume is bounded by the surface of the soil and the depth of the root zone. For full cover crops, such as surface and sprinkler irrigation, the system is one dimensional. For crops in which the rootzone is partially wetted, such as drip irrigation, the system must be considered as two or three dimensional.

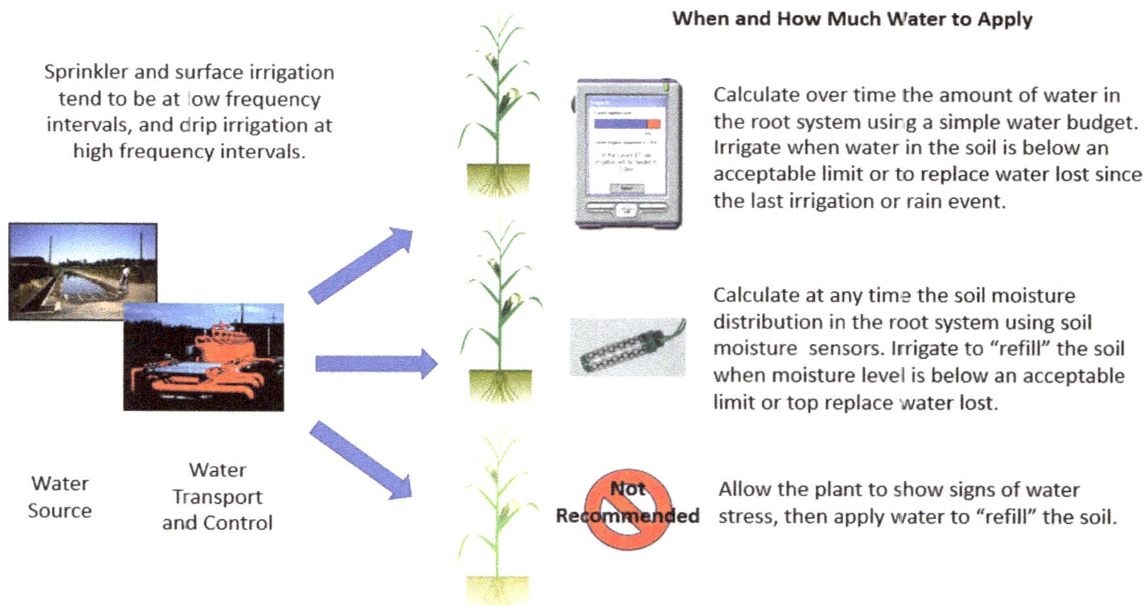

Figure 4 Example strategies can be used to schedule irrigation in agriculture. The frequency is dependent on the type or irrigation system used and the amount of the soil moisture distribution at the time of irrigation.

In fluid dynamics, the mass conservation equation is given in integral form as:

$$\frac{dq}{dt} + \oiint_S j\,dS = 0 \tag{1}$$

where $\boldsymbol{S}$ is a surface vector that encloses a control volume (L^2), q is the amount of mass (or volume of water) in the control volume (L^3/T), j is a vector that defines the velocity field at the surface $\boldsymbol{S}$ (L/T), and t is time (T).

In simple terms, equation 1 states that the change in the amount of water inside a control volume is equal to the net sum of the amount of water transfer across the boundary. That is, for any given period, the volumetric change of water stored in the soil root system of an irrigated crop is the result of the difference between water entering the control volume and water exiting the control volume. This is often referred to in the irrigation literature as the water budget. The simplest version of this budget assumes that the control volume is the volume consisting of the cropped surface area with a depth equal to the root depth of the plant capable of extracting water, and that water transfer occurs only across the control volume's surface. Thus,

$$\Delta S = Ir + Re - ETc - D \tag{2}$$

where ΔS is the change in stored volume of water stored over a given time interval (L^3/L^2), Ir is the net volume of water that is applied by irrigation to the cropped area (L^3/L^2), Re is the amount of water from precipitation that effectively enters the soil (L^3/L^2), ETc is evapotranspiration of the crop, the combined evaporation from the soil, and transpiration by plants (L^3/L^2), and D is the volume of water leaving the system by deep percolation (L^3/L^2). Note that the stored volume of water cannot exceed the capacity of the control volume (Fig. 5).

Because early irrigation systems, such as surface and sprinkler systems, fully cover the cropped area, the values in equation 2 are often presented as volume of water applied per unit area or depth. This simple equation has been used as the basis to determine the amount and timeliness of irrigation for many years (Weren, 1978), well before computer-based tools became available. As other technologies emerged, such as drip and microsprinkler irrigation, that do not irrigate 100% of the cropped area but irrigate a fraction of the root zone, variations of the above equation emerged that partitioned the irrigated zone from the non-irrigated zone, that is, the root zone that received water from irrigation and rainfall, and the zone that received water only from rainfall. Particularly in humid regions, the contributions of rainfall outside of the irrigated root zone are important (Phene and Beale, 1976; Parsons and Wheaton, 2000).

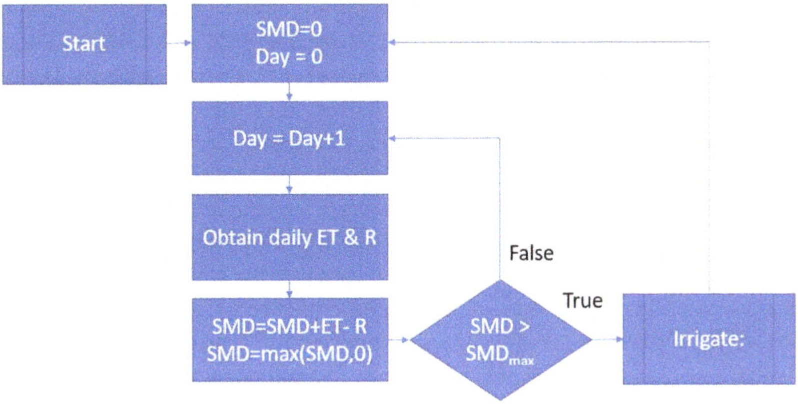

Figure 5 The checkbook irrigation scheduling algorithm assumes knowledge of the usable amount of water in the soil, usually a percentage of the field capacity (SMD_{max}). After heavy rain or irrigation, daily ET is added, and rainfall subtracted to this amount to obtain the water deficit in the soil on any given following day. When the available amount of water reaches a minimum allowable value, that is, SMD_{max}, irrigation is applied bringing the system to field capacity or zero deficit. Note that an additional volume of irrigation is applied to maintain a salt balance. Also, the required irrigation volume is corrected to account for the efficiency of the system.

5.1.1 Low-frequency irrigation

One strategy used in the application of this equation is to assume that the control volume can hold a maximum volume of water, usually field capacity (FC) and that any volume of water above this volume entering the system is lost through drainage and/or runoff. Second, there is a minimum acceptable value of water in the soil that ensures the plant does not suffer hydric stress. The difference between FC and the minimum acceptable value for water in the soil is represented by an allowable soil moisture deficit (SMD_{max}) given as a percentage of FC and is the fraction of usable water in the soil.

When irrigation or rainfall occurs, water in the soil above FC water exits through drainage and runoff, thus FC is the maximum value for water in the soil. Over time, water is lost via ET and gained from rainfall until the depletion level of water in the soil triggers an irrigation event.

When irrigation is applied, an amount of water above that needed to meet the difference between irrigation and rainfall that infiltrates into the soil is applied to ensure that salts do not accumulate in the soil. This additional amount of water has been expressed as the leaching requirement (LR). A large body of research focused on assessing salt tolerance (Maas and Hoffman, 1977). Although this is common in arid regions, in humid regions, rainfall may be sufficient to flush salts from the root system.

This method is simple enough that it is widely used for irrigation scheduling. Thus, in a fully cropped area where the allowable SMD is known and daily rainfall and evapotranspiration data are available, a simple water budgeting procedure can be implemented, often referred to as the 'checkbook method' for its parallel to deposits and withdrawals to a bank account. Figure 5 shows the basic steps followed in implementing this method, which can easily be done manually and often appears in extension fact sheets and bulletins of the Agricultural Extension Service (Weren, 1978). Since this method oversimplifies a complex dynamic system, it is recommended that the checkbook method for irrigation scheduling be used with complementary tools, such as soil moisture sensors, plant stress observation, canopy temperature, sap flow measurements, or other indicators of the status of water in the soil or plant-water stress, thus starting at an event that raises the soil to FC (irrigation or rainfall), on a daily time step *t*:

$$\Delta S = \sum_{t_j - 1}^{t_n} (ETc_{ti} - R_{ti}) \tag{3}$$

When the *S* reaches the allowable depletion, then water is applied to raise the soil to FC and drain salts out of the system:

$$Ir = E \frac{\Delta S}{\left(1 - {EC_i}/{EC_d}\right)} \tag{4}$$

where *E* is the irrigation system's efficiency, EC_i is the electrical conductivity of the irrigation water ($M^{-1} L^{-2} T^3 A^2$), and EC_d is the electrical conductivity of the water draining below the root zone ($M^{-1} L^{-2} T^3 A^2$). Equation 4 is based on a simple assumption of maintaining a salt balance and is a critical issue for salt-affected areas or areas with saline irrigation water (Hopmans et al., 2021).

For a well-designed system, irrigation efficiency is dependent on accurate estimates of *ETc* and *R*. A substantial body of knowledge has been accumulated over many years of research to determine *ETc*. The most widely accepted is the FAO56 and the dual Kc approach (Pereira et al., 2020). Also note that at any time, *R* is the fraction of rainfall remaining in the root zone, that is, *R* cannot exceed the water deficit at the time of a rainfall event.

5.1.2 High-frequency irrigation

A second irrigation strategy is that of high-frequency irrigation. This approach resulted from the development of irrigation systems capable of delivering water to the soil frequently and in small quantities (Rawlins, 1973), such as drip and microsprinklers. As opposed to low-frequency irrigation where the irrigation cycle consists of a brief period of infiltration followed by an extended period where the crop uptakes water from the soil, high-frequency irrigation applies water in short, frequent intervals to maintain a relatively constant environment within the root system (Howell and Meron, 2007). Thus, a typical implementation of this approach is to irrigate daily to 'replenish' ET of the previous day. This approach can be implemented using a water budget. However, these systems partially wet the root zone of the crop. The volume of the root zone irrigated by the system becomes crucial, as it is important that it be of sufficient size for the roots to supply water to the plant, impacting the necessary number of emitters, which is an important design constraint for any irrigation system that depends on the soil and the crop.

5.2 Soil moisture measurement

An alternate approach to the irrigation scheduling is the nondestructive measurement of soil moisture distribution using soil moisture sensors. In general, an ideal soil moisture sensor is a device that produces a signal that varies monotonically with soil moisture and is unaffected by other system properties, such as salinity and temperature. Reviews of soil moisture sensors were conducted by Zazueta and Xin (1994), Garg et al. (2016), and Yu et al. (2021). Two strategies are used when sensors are used with irrigation systems. One strategy consists of using the sensors to establish the current moisture distribution and apply the necessary water to bring soil moisture to a desired level, that is:

$$V_w = \iiint_{CV} \left(q_d\left(x,y,z\right) - q\left(x,y,x\right)\right) dxdydz \tag{5}$$

where V_w is the volume of water to apply (L^3), Θ_d is the desired soil moisture distribution (L^3/L^3), and Θ is the actual soil moisture distribution (L^3/L^3) in the control volume.

A second strategy is to use one or more soil moisture sensors in such a way that if a threshold of soil moisture is reached, an irrigation event is triggered (Smajstrla and Koo, 1986; Muñoz-Carpena et al., 2008). Figure 6 shows an example of this approach. Using sensors involves several decisions that will affect irrigation efficiency and the performance of the system, in particular the number of sensors or blocks of sensors, their location in the field, the depth(s) of installation, and location near the plant, and in addition the effect of variables other than that being measured, maintenance requirements, and cost.

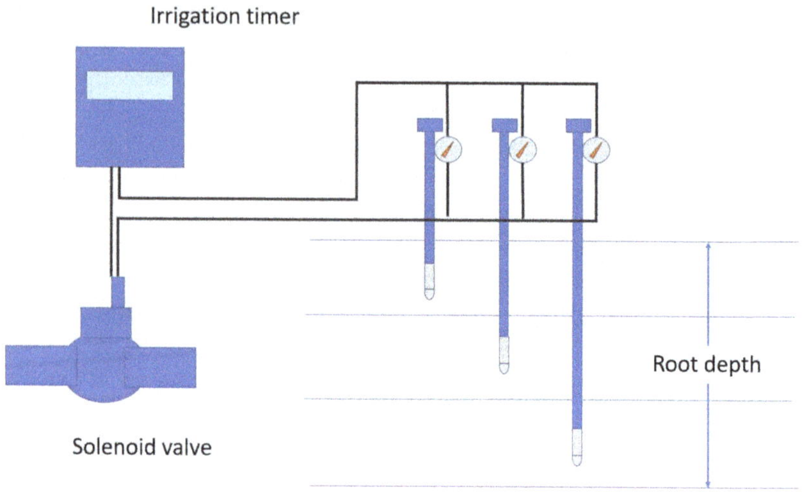

Figure 6 A simple self-adjusting automated irrigation controller can be constructed with sensors that will close a circuit when a low-moisture threshold is reached. In this example switching tensiometers are used, although any threshold sensor can be used. The irrigation controller is set to short high-frequency periods of irrigation. When any one of the sensors reaches the low soil water tension threshold, they will close the circuit, allowing power to the solenoid valve and applying irrigation. The irrigation applied is a small amount allowing for water to redistribute itself before the next irrigation attempt by the controller. Note that tensiometers will only work well on soils such that the threshold is within the operating range of the instrument.

5.3 Water movement prediction based on soil dynamics

Equations describing mass and heat transfer under moisture and temperature gradients were developed by Philip and de Vries (1957). Theoretical approaches to describe the movement of solutes (Jury, 1982) or bacteria (White et al., 1986) and the governing equations that describe the simultaneous transfer of heat, water, and the soil (Nassar and Horton, 1992; Peters et al., 2019) were developed. The fundamental questions of when and how much to apply can be addressed using the equation that describes dynamic unsaturated flow in an isothermal, homogeneous, isotropic soil (Richards, 1931).

$$\nabla\left(k\nabla h\right)+\frac{\partial k}{\partial z}+S=\frac{\partial \theta}{\partial t} \tag{6}$$

where Θ is soil moisture content (L^3L^{-3}), k is hydraulic conductivity (L/T), h is hydraulic pressure head (L), z is elevation (L), S is a sink term accounting for root absorption of water ($L^3L^{-3}T^{-1}$), and t is time (T).

Because S is a function of h and space, and k and Θ are functions of h, equation 5 is highly nonlinear. As a result, analytical solutions to this differential equation are available only for simplified conditions. However, the application of analytical solutions provides insights that are useful for design and to study the behavior of the system, particularly in systems, such as drip and microirrigation, that cannot be reduced to one dimension. Analytical solutions of equation 5 are discussed in detail by Ravi et al. (1998), including steady-state solutions, quasi-linear approximation solutions, and transient infiltration solutions. Applicable to a broader range of problems than those addressed by analytical methods are semi-analytical models, some of which predict the movement of solutes under plant uptake in the soil (Mmolawa and Or, 2000).

Because of the nonlinearity of equation 5 and the boundary conditions found under drip irrigation, many numerical algorithms have been presented, including finite difference and finite element methods. These methods have the advantage that they can be used in nonhomogeneous isotropic media, can handle boundary conditions and sources and sinks, and fit irregular geometries of the control volume (Zazueta et al., 1985). A comprehensive review of models for predicting soil water dynamics is presented by Subbaiah (2013).

Another approach to modeling infiltration and distribution of water in the soil is to use simplified physical approaches, such as the Green–Ampt equation (Green and Ampt, 1911), that models one-dimensional infiltration into a soil if a saturated front moves into the soil driven by gravity and capillary action. This approach was extended to a three-dimensional model for infiltration into a soil from a point source by Chu (1994). This model was further extended to include the infiltration phase and calculate the three-dimensional soil water redistribution phase between point-source events (Gowdish and Muñoz-Carpena, 2018).

There are several empirical and semiempirical methods that allow estimations of wetted bulb diameters, usually narrowly limited to the conditions under which they were developed. This includes reports from field tests (Goldberg et al., 1976), empirical equations (Schwarzman and Zur, 1986; Zazueta, 1992), and tables (Keller and Bliesner, 1990). Although these are useful in determining the size of the wetted bulb to determine the number of emitters required to wet a given fraction of the root zone during the design stage of the irrigation system, these are not useful for scheduling purposes.

5.4 Plant water stress indicators

Plants are complex organisms that react to changes in the environment. Thus, the response of plants that are sensitive to water stress can be used as a tool in irrigation. Plant response accounts for the cumulative effect of stress due to declining soil moisture status and the increase in evaporative demand of the atmosphere that results in an observable water stress level in plant. Understanding the biology of the plant at a level that can be used to make

Table 2 Plant-based water stress indicators

Leaf water potential (LWP)	LWP is the force that drives transport of water from roots to the leaves.
Relative water content (RWC)	RWC is a comparison of the water content of a leaf to the maximum possible water content of the same leaf at its full turgor.
Photosynthesis	Photosynthesis is the key factor affecting biomass accumulation and economic yield. Under water-stressed conditions, overall photosynthesis decreases due to decrease in both leaf area and photosynthetic rate per unit of leaf area.
Stomatal conductance (SC)	SC is a measure of the degree of physical resistance to gas movement between air and leaf interior.
Canopy temperature (CT)	CT affects the metabolic activities of a plant. Canopy temperature is regulated by transpiration.
Hormonal balance (HB)	HB is affected during low soil moisture availability. Root cells start dehydrating due to low water uptake by the cells. When dehydrating, root cells start synthesizing abscisic acid (ABA). This ABA is then transported to leaf apoplast via xylem vessels, resulting in complex physiological and biochemical plant adjustments.
Osmotic adjustments (OA)	OA is the process of accumulation of osmotically active and metabolically neutral solutes, such as amino acids (primarily proline and glycine), organics acids, sugars, betaine, and polyamines. Accumulation of solutes results in lower osmotic potential, which is crucial for maintaining the water status and physiological activity of plant cells under water stress.
Electrolyte leakage (EL)	EL is an indirect measurement of the integrity and stability of cell membrane under stress conditions.
Sap flow (SF)	Sap flow represents the transpiration flow as the ascent of sap within the xylem tissue. Under water-stressed conditions, transpiration rate decreases due to stomatal closure, which subsequently affects the sap flow.
Stem and fruit diameter (SFD)	SFD varies in response to movement of water between the xylem and phloem.
Leaf expansion (LE)	LE expansion rate is affected by leaf temperature, photon flux density, evaporative demand, and soil water content. Generally, loss in leaf turgor under water stress results in reduced cell expansion.

water management decisions and the technical solutions to measure the related variables can be used to optimize the value returned from water-related investments. While the effects on plants of water deficit continue to be the subject of study, there are several responses that can serve as water stress plant-based indicators, as shown in Table 2 (Parkash and Singh, 2020). Particularly in areas in which water is scarce or becoming scarce due to climate change, demographic pressures, or increases in purchasing capacity, it is important to apply water when reaching a deficit level that will not have a detrimental effect on the value of the crop. Thus, a strategy for water use (deficit irrigation) is to optimize the productivity of water by focusing on drought-sensitive growth stages of a crop and restricting water during drought-tolerant phenological stages (Geerts and Raes, 2009) (Sun et al., 2020). The advantages and disadvantages of various approaches were discussed by Jones (2004, 2008) and Ihuona and Madramootoo (2017).

5.5 Other approaches

Statistical methods. Using historical records to simulate the behavior of an irrigation system over a long period of time can provide valuable insights related to irrigation scheduling. Zazueta et al. (1988) used a 25-year weather record to normalize the temporal and spatial distribution of irrigation intervals and condense the results to simple recommendations. Parsons and Morgan (2022) produced tables to schedule irrigation of Florida citrus. Although, in practice, this approach is not recommended for scheduling irrigation in large agricultural operations, it is useful for small irrigation systems where efficiencies

Table 3 Machine learning and evolutionary algorithms used in the estimation of evapotranspiration

ANFIS	Adaptive Neuron Fuzzy Inference System
ANN	Artificial Neural Network
ANN-GWO	Artificial Neural Network-Gray Wolf Optimization
CART	Classification and Regression Tree
CCNN	Cascade Correlation Neural Network
ELM	Extreme Machine Learning
FL	Fuzzy Logic
GEP	Gene Expression Programming
GEP	Gene Expression Programming
GP	Genetic Programming
GRNN	Generalized Regression Neural Network
KNN	K Nearest Neighbor
LS-SVR	Least Squares - Support Vector Regression
MLP	Multilayer Perceptron
OPELM	Optimally Pruned Extreme Learning Machine
OSELM	Online Sequential Extreme Learning Machine
RF	Random Forest
SELM	Sequential Extreme Learning Machine
SVM	Support Vector Machine
SVM-FFA	Support Vector Machine - Firefly Algorithm
SVR	Support Vector Regression
WNN	Wavelet Neural Networks
XGBoost	Extreme Gradient Boosting

Table 4 Research on ML and AI for the estimation of daily evapotranspiration and evaporation

Reference	Estimated parameter	Algorithm(s) used
Wu et al. (2021)	ETo	ELM, MLP, RF, SVRGRNN, XGBoost
Bai et al. (2021)	ETo	RF, KNN, SVM, MLP
Heddam et al. (2020)	ETo	OSELM, OPELM
Maroufpoor et al. (2020)	ETo	ANN-GWO, LS-SVR
Falamarzi et al. (2014)	ETo	ANN WNN
Goyal et al. (2014)	Daily Pan Ev	ANN, ANFIS, LS-SVR, FL
Gocić et al. (2015)	ETo	GP, SVM-FFA,ANN, SVM Wavelet
Dou and Yang (2018)	ETo	ELM, ANN, SVM, ANFIS
Saggi and Jain (2019)	ETo	H2O-DL, RF, GBM. GLM
Saggi et al. (2022)	ETo	ELM, MLP, SVM
Sharma et al. (2021)	ETo	CNN-LSTM, Conv-LSTM
Bellido-Jiménez et al. (2021)	ETo	MLP, ELM, GRNN, SVM, RF and XGBoost
Yaseen et al. (2020)	Daily Pan Ev	CART, CCNN, GEP

Source: Adapted from: Saggi et al. (2022).

are low due to mismanagement. In addition, the expectation that the weather will behave in the future as it has behaved in the past is questionable.

Fuzzy logic. Because at many locations, data required to make irrigation-related decisions are partially known (Huang, et al., 2010), fuzzy theory has been proposed for irrigation scheduling (Patel et al., 2012) and automation of irrigation systems (Krishnan et al., 2020).

Neural networks (NN) and machine learning (ML). The estimation of daily evapotranspiration is crucial for irrigation scheduling as higher time intervals do not have the resolution necessary to make adequate scheduling decisions. Recently, a substantial amount of work using NN and ML has been focused on estimating daily reference evapotranspiration and crop evapotranspiration, as well as managing irrigation from soil sensor data. Table 3 summarizes different approaches to the determination of daily evapotranspiration. Table 4 refers to recent research and developments in the use of these technologies.

6 Examples of decision support systems over the years

Early DSS tools relied primarily on a stand-alone PC. Later, these evolved to acquire data from remote sites. With the appearance of the public Internet, tools appeared with a user interface based on a browser. At the same time, computers evolved to handheld devices and eventually intelligent phones, resulting in the contemporary app, that is, a DSS with a user interface that meets the requirements of being correct, usable, convenient, and efficient.

Today, advances are being driven by developments in technology such as high-performance computing, IoT, big data, analytics, and AI. The immediate future anticipates practical uses of quantum computing.

6.1 The first generation

The first generation of practical computer-based DSSs was the result of the wide adoption of the PC and the related stand-alone applications. Some examples follow.

Spreadsheets. Business applications played an important role in the adoption of computer-based DSSs in agriculture. Starting with simplistic recordkeeping and spreadsheets, business operations simplified management of the agricultural enterprise (Bricklin, 2016). The potential impact of these machines was immediately recognized by the agricultural community (Bottcher and Zazueta, 1983; Weber et al., 1983). By the end of the first decade, from the appearance of the PC, these applications became pervasive. An early application of spreadsheets to irrigation was carrying out a water balance to make decisions about irrigation. Farmers were taught in computer workshops to use spreadsheet water budgeting. The transition from the checkbook method (Weren, 1978) to a spreadsheet was a natural transition for first-time users of computer-based DSS tools. This method, when used with other indicators of plant water stress, was a useful tool in managing irrigation applications.

Stand-alone software applications. Applications dedicated to a single purpose were developed for agriculture when the PC became widely available since the early 1980s. These applications included a wide range of subjects related to water management on the farm, including but not limited to hydraulic design of irrigation systems, ET estimation methods, scheduling, fertilizer injection, water amendments, automation, energy, etc. Many of these applications do not exist today primarily due to obsolescence (e.g. improvements in knowledge or evolution of strategy used) or changes in the computer platform (e.g. the operating system, interpreters, and compilers). However, more useful applications emerged to replace them taking advantage of newer technologies.

Expert systems (ES). An ES is a computer program that solves problems in a narrow knowledge domain with a performance equal to or better than a human expert. The systems are built through a knowledge elicitation process where a knowledge engineer extracts the rationale from a domain expert, in the form of rules (if *antecedent* then *consequent*) by which an expert or a group of experts address a problem. Using these rules, an inference engine and input provided by a user can lead to an actionable recommendation to solve a problem (Fig. 7). Expert systems were applied to different aspects of irrigation management, including scheduling, evaluation, crop selection, equipment selection, and problem diagnosis (Mohan and Arumugam, 1997;

Figure 7 Drip irrigation emitters clogged with iron bacteria action deposits. The presence of *Pseudomonas* and *Enterobacter* and small concentrations (ppm) of iron or magnesium result in bacterial growth that can clog a large conduit. Proper diagnosis as well as a prescriptive solution to solve the problem are required. Because of the interaction complexity of the many factors leading to this condition, a DSS in the form of an expert system was developed (Zazueta, 1983) based on heuristics (Ford, 1979).

Bwambale et al., 2022). In addition, these were used not only for diagnosis and treatment but also were applied to real-time management of irrigation systems (Xin and Zazueta, 1995). Expert systems are based on rules, and as such, they can explain the reasoning that leads to a given conclusion. However, although ES continues to be developed, interest in ES diminished substantially due to the difficulty and effort required in the knowledge elicitation stage, the difficulty to maintain, and the emergence of alternate AI technologies. Expert systems use today is often associated with other solutions that involve different crop production subsystems (Lu et al., 2022).

Modeling and simulation. A model in this context is a program that behaves like a real system. A simulation is the operation of a model under a given set of conditions, most often to predict the behavior of the real system. Models are extremely useful as they provide insights into system behavior that is too expensive or risky to ascertain from a real system or because the timescale of the phenomenon being studied is too large. Models often used in agriculture are a combination of physical, chemical, and biological knowledge, as a result, they are a powerful tool to assist in decision-making. Modeling and simulation as tools for research, design, and management of irrigation systems are abundant in the literature. Pereira et al. (2020) reviewed 32 models based on FAO56 Average and dual Kc for the estimation of ETC in the water balance going back to 1999. From the early applications of modeling and simulation in agriculture, these were used as tools to manage on-farm activities, such as irrigation (Zazueta and Lokers, 1991; Zazueta, 1993), climate (Jagtap et al., 2002), farm machinery operations (Busato et al., 2013), plant and animal housing environments (Singh, 2022), predicting crop behavior and yield for policy decisions (DSSAT Foundation, 2022), and many other applications. In general, DSSs have evolved based on the same basic principles and have been adapted and improved over time as better knowledge is acquired about the behavior of an irrigated system and as new technology developments become available.

Information delivery systems. An important factor for successful agriculture requires knowledge that can be applied at the farm level. Although technically, information delivery systems are not DSSs, access to information has played a significant role in improving agricultural outcomes and reducing environmental impact. Modern agriculture's success is to a large degree to access to actualized and relevant information in the farm. Mechanisms for delivery of knowledge vary from country to country. One such mechanism that proved successful in the past is the agricultural extension or the equivalent organization charged with delivering relevant science-based information to agricultural stakeholders. Prior to the advent of electronic means, one of the main mechanisms for delivery of information to farmers was the Extension Fact Sheet (or equivalent), which consisted of condensed practical information addressing, among other issues, management of agricultural systems. With the public Internet, a transformation in the delivery of information occurred, with information delivered faster, at any location, and on many devices. Ironically, the issue now is not a lack of information, but the volume, and uncertainty of the quality of the information available.

6.2 The second generation

As the use of IT in agriculture became common place, its applications became more complex and useful. The main driver was networking, and the simplicity of use of the Internet browser, which allowed moving data quickly to and from the point of use (Greenstein, 2015). Examples of these applications follow.

Web-based irrigation scheduling. As access to remote information became available and agricultural weather data acquisition networks became widely available applications shifted to the web. For example, the web irrigation scheduling tools in North Dakota (Schrerer, n.d.), and Florida (Morgan, n.d.).

Also, with the developments in small, connected computers, numerous DSSs were developed for handheld PCs (Sepúlveda et al., 2005) and smartphones (Bayabil et al., 2020).

Agricultural weather networks. Weather information is critical for management of agricultural operations. Weather information has consistently been among the services that are provided to agricultural stakeholders via electronic means. Before PC became available, mainframe systems were used with connections by modem to data centers where farmers could download this information. Today, these networks have evolved to specialized equipment networked via wireless connections that collect comprehensive weather information and present in a way that is useful to agricultural stakeholders. Accurate estimation of ETo requires instrumentation that is not typically available to a farmer. Thus, having a reliable and accurate estimation of ETo for water budgeting made it convenient and possible to estimate the amount and timeliness of irrigation with higher accuracy. In addition, climate predictions made it possible for farmers to strategically plan for water management-related activities.

6.3 The third generation

As computers became widely adopted and more common in the farm different types of applications began to share or provide data for other applications. As a result, integration of various systems began to appear. An example of these systems is the SICODE system that was used for management of water resources in Module 3 of Irrigation District 05, Mexico (Mundo et al., 1997). The Irrigation district is in Delicia's Chihuahua, Mexico. It compromises 80 102 ha and provides water to 9650 users (farms). The system was developed to improve water delivery to users. It relied on a geographical information system (GIS) that included the location of water sources, canal infrastructure, and points of delivery. In addition, the GIS was used to record the location of the different crops at each plot grown in the district. A water balance was conducted for every plot in the irrigation district based on data from an automated station, the soil, and the crop at each location. This water balance was used to determine water use and needs at different points of delivery. With this information, an expert system established priorities for irrigation water delivery using the open-channel infrastructure.

As IT and related technologies became more sophisticated and powerful many areas of agriculture were affected and farm management thinking became more complex, as shown by Sørensen et al. (2010) and Arvanitis and Symeonaki (2020). In the domain of agricultural and biosystems engineering, advances were made in automation. For example, the use of computers to control the growth environment of the plant through precise control of environmental factors such as water, nutrition, the atmosphere, plant selection, artificial soils, and disease control. Examples of these are modern greenhouses (Shamshiri et al., 2018; Marcelis et al., 2019) and plant factories (Avgoustaki and Xydis, 2020). Satellite-based positioning made geographical information systems (GIS) possible, which after being coupled with sensors, actuators, and databases gave rise to precision agriculture. Not only were systems that affected field operations and management developed and widely applied, but also intelligent machines used in related postharvest technologies and food engineering with a trend to integrate information across the production and supply chain. As a result, DSS are becoming more complex requiring more effort on part of the users. However, it appears that users are willing to expend the extra effort resulting from complexity if there are gains in the value of using the tool (Chenoweth et al., 2004).

6.4 The fourth generation

Advances in technology in the recent past have the potential to transform agriculture in many ways (Klerkx and Rose, 2020). These are driven by the euphemistically called fourth Industrial Revolution (4IR) Technologies (see Fig. 8).

Massive amounts of data. As technology has become pervasive in the agri-food chain, the amount of information has given rise to large quantities of data. In this chain - that is really a network - mass, energy, value, and information are transferred across the system generating data in very large amounts, of many different types, and often of dubious quality (Abderahman et al., 2022). This 'Big Data' requires new mathematical and statistical tools for its interpretation and analysis to yield useful information. That is, new analytics, or the collection of tools to describe, predict, and prescribe solutions in domains ranging from

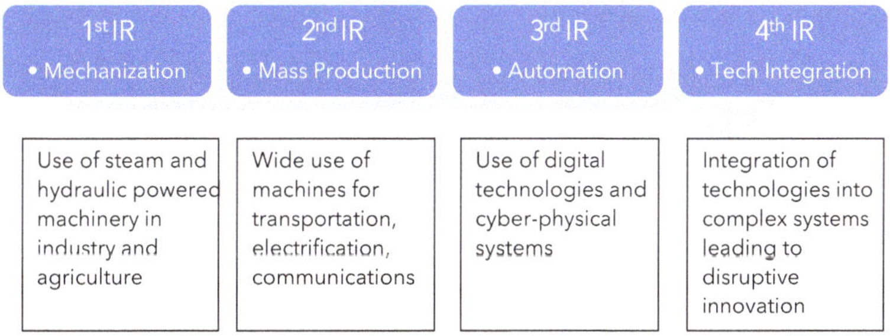

Figure 8 Starting with the invention of the steam engine and hydraulic power, the world has seen deep transformations driven by technology. From the second half of the nineteenth century and first half of the twentieth century, technologies such as the automobile and the lightbulb changed the way we work and connect with other humans. Also, because of its wide impact, other domains became important, such as government intervention to create a national electric grid and a highway network for transportation, each with its many ramifications, economic, legal, environmental, etc. The impact of recent technological developments is expected to bring about disruptive change (Barret and Rose, 2020; World Economic Forum, 2022).

field operations to national policy, is necessary to realize new opportunities and assess risks. While many DSS will remain tactical to address issues local to the farm, we are likely to see DSS that will tend to be strategic in nature, incorporating farm operations to the broader supply chain.

Hyperintelligent machines. For decades, researchers and developers continue to improve the level of intelligence of machinery. Not only are machines programmable but they can make decisions under complex situations, extract knowledge from data sets, and learn. Machines are evolving in such a way that not only will they operate individually, but collectively will share knowledge, within and beyond the farm to make coherent group decisions for tactics aligned with a predefined strategy. As the capacity and intelligence of devices and machinery used in agriculture improve, machines become networked: they collaborate with each other and humans to achieve a given task. Beginning with 'swarms' of robots for agricultural operations (Albiero et al., 2022). Furthermore, with 4IR technologies, it is now foreseeable that systems can collaborate with each other across the supply chain (Brauna et al., 2018) connecting networks of networks.

Complex systems. A modern approach to systems thinking in agriculture and food production uses complex system theory (Ladyman et al., 2013). It is now widely recognized that solutions to the problem of providing wholesome food in the necessary quantity and quality cannot be solved through narrow engineering solutions. These involve multiscale (spatial and temporal) and multidisciplinary interdependencies that include factors such as understanding the behavior of parts of a system but its whole, nonlinearity in cause-effect, self-organization, and complex hierarchical organizations among others. Nevertheless, IT plays, and will continue to play a major role in the implementation of solutions for complex systems. As IR4 technologies (Table 1) become pervasive, they will continue to be transformative (Kabir, 2019).

7 Conclusion

Tools to aid managers to improve decision outcomes have evolved closely with available technology. Currently, technology developments up to the first quarter of the twenty-first century have focused in developing solutions for narrow problems. It is now clear that there is an evolution in the adoption of technologies, in that these are converging to create new opportunities and are likely to have deep economical, societal, and personal implications. Addressing the difficulties presented by current and emerging problems requires a complex system approach. For the developer of DSS, this poses a challenge in producing useful tools that are tailored to the competencies of different audiences while hiding the complexity associated with a diagnostic

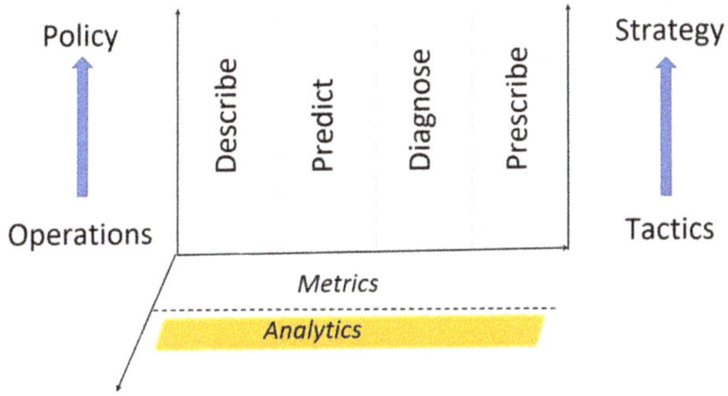

Figure 9 Agriculture production and related activities to deliver goods produce large amounts of data. Metrics are key data that gauge performance, while analytics is the use of these metrics to make decisions. These can be used at many different levels by different stakeholders, from decision support systems that help farm operations, to policy and strategy decisions. For example, metrics and analytics can be used to describe a problem, predict what the outcome will be of different actions, and most importantly, prescribe a solution.

and prescription of a problem in this context. As developments in technology take place, DSS are likely to expand from narrow problem-solvers to the domain of strategy and policymaking (Fig. 9). Especially in the domain of using water resources in the farm, the competing demands for water will require innovation to maximize the value to society of irrigated agriculture.

8 Where to look for further information

Tools for improved decision making are available from a variety of sources. These include insights from scientific journals on the subject as well as those provided by extension and outreach organizations. Examples of these include the *Journal of Irrigation and Drainage Engineering* (https://ascelibrary.org/journal/jidedh), and the *Extension Digital Information Source* (https://edis.ifas.ufl.edu/). In addition, irrigation equipment manufacturers often provide tolls specifically tailored to their products.

9 References

Abderahman, R., Keog, J. G. and Rejeb, K. (2022). Big data in the food supply chain: a literature review. *Journal of Data, Information and Management* 4(1), 33–47. doi:10.1007/s42488-021-00064-0.

Albiero, D., Garcia, A. P., Umezu, C. K. and de Paulo, R. L. (2022). Swarm robots in mechanized agricultural operations: a review about challenges for research. *Computers and Electronics in Agriculture* 193, 1–17. doi:10.1016/j.compag.2021.106608.

Alcamo, J., Flörke, M. and Marker, M. (2010). Future long-term changes in global water resources driven by socio-economic and climatic changes. *Hydrological Sciences Journal* 52(2), 246–275. doi:10.1623/hysj.52.2.247.

Arvanitis, K. G. and Symeonaki, E. G. (2020). Agriculture 4.0: the role of innovative smart technologies towards sustainable farm management. *Open Agriculture Journal* 14(1), 130–135. doi:10.2174/1874331502014010130.

Avgoustaki, D. D. and Xydis, G. (2020). Plant factories in the water-food-energy Nexus era: a systematic bibliographical review. *Food Security* 12(2), 253–268. doi:10.1007/s12571-019-01003-z.

Bai, Y., Zhang, S., Bhattarai, N., Mallick, K., Liu, Q., Tang, L., Im, J., Guo, L. and Zhang, J. (2021). On the use of machine learning based ensemble approaches to improve evapotranspiration estimates from croplands across a wide environmental gradient. *Agricultural and Forest Meteorology* 298–299, 108308. doi:10.1016/j.agrformet.2020.108308.

Barret, H. and Rose, D. (2020). Perceptions of the fourth agricultural revolution: what's in, what's out, and what consequences are anticipated? *Sociologia Ruralis* 62(2), 162–189.

Bayabil, H., Migliaccio, K., Debastiani Andreis, J., Fraisse, C., Morgan, K. and Vellidis, G. (2020). *Smart Irrigation Apps: Urban Turf*. Gainesville, FL: UF IFAS Extension. Retrieved from https://edis.ifas.ufl.edu/pdf/AE/AE49900.pdf (accessed September 7, 2022).

Bellido-Jiménez, J. A., Estévez, J. and García-Marín, A. P. (2021). New machine learning approaches to improve reference evapotranspiration estimates using intra-daily temperature-based variables in a semi-arid region of Spain. *Agricultural Water Management* 245, 106558. doi:10.1016/j.agwat.2020.106558.

Bottcher, D. and Zazueta, F. (1983). Proceedings of the of the 1st International Computer Conference on Agricultural Extension Programs. 1st International Computer Conference on Agricultural Extension Programs (p. 874). Tampa: Institute of Food and Agricultural Sciences, University of Florida.

Bozkurt, S. and Ozekici, B. O. (2006). Effects of fertigation managements on clogging of in-line emitters. *Journal of Applied Sciences* 6(15), 3026–3034. doi:10.3923/jas.2006.3026.3034.

Brajović, M., Vujović, S. and Đukanović, S. (2015). An overview of smart irrigation software. 4th Mediterranean Conference on Embedded Computing (MECO) (pp. 353–356). doi:10.1109/MECO.2015.7181942.

Brauna, A.-T., Colangelob, E. and Steckel, T. (2018). Farming in the era of industries 4.0. 51st CIRP Conference on Manufacturing Systems (pp. 979–984). Elsevier. doi:10.1016/j.procir.2018.03.176.

Bricklin, D. (2016, November). Meet the inventor of the electronic spreadsheet. Retrieved from Ted.com: https://www.ted.com/talks/dan_bricklin_meet_the_inventor_of_the_electronic_spreadsheet?language=en (accessed April 9, 2023).

Busato, P., Sørensen, C. G., Pavlou, D., Bochtis, D. D., Berruto, R. and Orfanoub, A. (2013). DSS tool for the implementation and operation of an umbilical system applying organic fertiliser. *Biosystems Engineering* 114(1), 9–20. doi:10.1016/j.biosystemseng.2012.08.012.

Bwambale, E., Abagalea, F. K. and Anornud, G. K. (2022). Smart irrigation monitoring and control strategies for improving water use efficiency in precision agriculture: a review. *Agricultural Water Management* 260, 107234. doi:10.1016/j.agwat.2021.107324.

Chenoweth, T., Dowling, K. and St. Louis, R. (2004). Convincing DSS users that complex models are worth the effort. *Decision Support Systems* 37(1), 71–82. doi:10.1016/S0167-9236(03)00005-8.

Chu, S. (1994). Green-aMPT analysis of wetting patterns for surface. *Journal of Irrigation and Drainage Engineering* 120(2), 414–421. doi:10.1061/(ASCE)0733-9437(1994)120:2(414).

Dou, X. and Yang, Y. (2018). Evapotranspiration estimation using four different machine learning approaches in different terrestrial ecosystems. *Computers and Electronics in Agriculture* 148, 95–106. doi:10.1016/j.compag.2018.03.010.

DSSAT Foundation (2022). Decision support system for agrotechnology transfer. Retrieved May 15, 2022, from DSSAT: https://dssat.net/

Evans, R. O. and Fausey, N. R. (1999). Effects of inadequate drainage on crop growth and yield. In: Skaggs, R. and van Schilfgaarde, J. (Eds). *Agricultural Drainage* (Vol. 38, pp. 13–54). ASA/CSSA. doi:10.2134/agronmonogr38.c2.

Falamarzi, Y., Palizdan, N., Huang, Y. F. and Lee, T. S. (2014). Estimating evapotranspiration from temperature and wind speed data using artificial and wavelet neural networks (WNNs). *Agricultural Water Management* 140, 26–36. doi:10.1016/j.agwat.2014.03.014.

Ford, H. (1979). Characteristics of slimes and ochre in drainage and irrigation systems. *Transaction SASE* 22(5), 1093–1096.

Garg, A., Munoth, P. and Goyal, R. (2016). Application of soil moisture sensors in agriculture: a review. Proceedings of the International Conference on Hydraulics, Water Resources and Coastal Engineering (Hydro2016) (pp. 1662–1672). Pune: CWPRS.

Geerts, S. and Raes, D. (2009). Deficit irrigation as an on-farm strategy to maximize crop water productivity in dry areas. *Agricultural Water Management* 96(9), 1275–1284. doi:10.1016/j.agwat.2009.04.009.

Gleick, P. H. (2009). A look at twenty-first century water resources development. *Water International* 25(1), 127–138. doi:10.1080/02508060008686804.

Gleik, P. H. (2018). Transitions to freshwater sustainability. *Proceedings of the National Academy of Sciences of the United States of America* 115(36), 8863-8871. doi:10.1073/pnas.1808893115.

Gocić, M., Motamedi, S., Shamshirband, S., Petković, D., Ch, S., Hashim, R. and Arif, M. (2015). Soft computing approaches for forecasting reference evapotranspiration. *Computers and Electronics in Agriculture* 113, 164–173. doi:10.1016/j.compag.2015.02.010.

Goldberg, D., Gornat, B. and Rimon, D. (1976). *Drip*. Kfar Shmaryhu: Drip Irrig. Sci. Pubs.

Gowdish, L. C. and Muñoz-Carpena, R. (2018). 3DMGAR: a transient quasi-3D point-source Green-Ampt infiltration and redistribution model. *Vadose Zone Journal* 17(1), 1–11. doi:10.2136/vzj2018.02.0032.

Goyal, M. K., Bharti, B., Quilty, J., Adamowski, J. and Pandey, A. (2014). Modeling of daily pan evaporation in sub-tropical climates using ANN, LS-SVR, Fuzzy Logic, and ANFIS. *Expert Systems with Applications*, 5267–5276. doi:10.1016/j.eswa.2014.02.047.

Green, W. and Ampt, G. (1911). Studies of soil physics, part I: the flow of air and water through soils. *Journal of Agricultural Science* 4, 1–24.

Greenstein, S. (2015). *How the Internet Became Commercial: Innovation, Privatization, and the Birth of a New Network*. Princeton, NJ: Princeton University Press. doi:10.1515/9781400874293.

Haman, D. (2022). Causes and prevention of emitter plugging in microirrigation systems. Retrieved May 5, 2022, from UF/IFAS Extension, University of Florida: https://edis.ifas.ufl.edu/publication/ae032

Heddam, S., Kisi, O., Sebbar, A., Houichi, L. and Djemili, L. (2020). New formulation for predicting daily reference evapotranspiration (et 0) in the Mediterranean region of Algeria country: optimally pruned extreme learning machine (OPELM) versus online sequential extreme learning machine (OSELM). In: Negm, A., Bouderbala, A., Chenchouni, H. and Barceló, D. (Eds). *Water Resources in Algeria-Part 1* (pp. 181-199). Springer International. Retrieved from https://www.springerprofessional.de/en/water-resources-in-algeria-part-i/18516262

Heermann, D. F., Martin, D. L., Jackson, R. D. and Stegman, E. C. (1990). Irrigation scheduling control and techniques. *Agronomy* 30, 509–535.

Hong, C. X. and Moorman, G. W. (2005). Plant pathogens in irrigation water: challenges and opportunities. *Critical Reviews in Plant Sciences* 24(3), 189–208. doi:10.1080/07352680591005838.

Hopmans, J., Qureshi, A. S., Kisekka, I., Munnsc, R., Grattan, S. R., Rengasamy, P., Ben-Gal, A., Assouline, S., Javaux, M., Minhas, P. S., Raats, P. A. C., Skaggs, T. H., Wang, G., De Jong van Lier, Q., Jiao, H., Lavado, R. S., Lazarovitch, N., Li, B. and Taleisnikp, E. (2021). Chapter One. Critical knowledge gaps and research priorities in global soil salinity. In: *Advances in Agronomy* (Vol. 169, pp. 1–191). doi:10.1016/bs.agron.2021.03.001.

Howell, T. A. and Meron, M. (2007). Irrigation scheduling. In: Lamm, F., Ayars, J. and Nakayama, F. (Eds). *Microirrigation for Crop Production: Design, Operation and Management* (Vol. 13, pp. 61–130). Elsevier. doi:10.1016/S0167-4137(07)80006-0.

Huang, Y., Lan, Y., Thompson, S., Fang, A., Hoffman, W. and Lacey, R. (2010). Development of soft computing and applications in agricultural and biological engineering. *Computers and Electronics in Agriculture* 71, 107–127. doi:10.1016/j.compag.2010.01.001.

Hutchinson, G., Carran, P. and McIndoe, I. (1993). IRRICAD-computerized irrigation design. In: Allen, R. G. *Management of Irrigation and Drainage Systems: Integrated Perspectives* (pp. 835–841). ASCE.

Ihuona, S. O. and Madramootoo, C. A. (2017). Recent advances in crop water stress detection. *Computers and Electronics in Agriculture* 141, 267–275. doi:10.1016/j.compag.2017.07.026.

Information Technology Task Force (2017). *ISO/IEC/IEEE 24765*:2017 *Systems and Software Engineering – Vocabulary*. International Standards Organization. Retrieved from https://www.iso.org/standard/71952.html (accessed December 6, 2021).

Jagtap, S., Jones, J., Hildebrand, P., Letson, D., O'Brien, J., Podesta, G., Zierden, D. and Zazueta, F. (2002). Responding to stakeholder's demands for climate information: from research to applications in Florida. *Agricultural Systems* 74(3), 415–430.

Jones, H. G. (2004). Irrigation scheduling: advantages and pitfalls of plant-based methods. *Journal of Experimental Botany*, 2427–2436. doi:10.1093/jxb/erh213.

Jones, H. G. (2008). Irrigation scheduling - comparison of soil, plant and atmosphere monitoring approaches. *Acta Horticulturae, 792*, 391–403. doi:10.17660/ActaHortic.2008.792.46.

Jury, W. A. (1982). Simulation of solute transport using a transfer function model. *Water Resources Research* 18(2), 363–368. doi:10.1029/WR018i002p00363.

Kabir, M. (2019). Technologies of the future. In: *Knowledge-Based Social Entrepreneurship* (pp. 91–133). Palgrave Studies in Democracy, Innovation, and Entrepreneurship for Growth. New York: Palgrave Macmillan. doi:10.1057/978-1-137-34809-8_4.

Keller, J. and Bliesner, R. (1990). *Sprinkle and Trickle Irrigation*. New York: Van Nostrand Reinhold.

Klerkx, L. and Rose, D. (2020). Dealing with the game-changing technologies of Agriculture 4.0: how do we. *Global Food Security* 24, 1–7. doi:10.1016/j.gfs.2019.100347.

Kneuper, R. (2018). *Software Processes and Life Cycle Models: An Introduction to Modelling, Using and Managing Agile, Plan-Driven and Hybrid Processes*. Cham: Springer. doi:10.1007/978-3-319-98845-0.

Knowx, P. and Griffin, M. (2022). ENSO impacts Southeast. Retrieved May 16, 2022, from Agriclimate.org: http://agroclimate.org/wp-content/uploads/2016/03/ENSO-Impacts-southeast.pdf

Köksal, Ö. and Tekinerdogan, B. (2019). Architecture design approach for IoT-based farm management information systems. *Precision Agriculture* 20(5), 926–958. doi:10.1007/s11119-018-09624-8

Krishnan, R. S., Julie, E. G., Robinson, Y. H., Raja, S., Kumar, R., Thong, P. H. and Son, L. H. (2020). Fuzzy logic based smart irrigation system using internet of things. *Journal of Cleaner Production* 252, 1–11. doi:10.1016/j.jclepro.2019.119902.

Kushwaha, D. and Misra, A. (2006). Cognitive Software Development Process and Associated Metrics - A Framework. 5th IEEE International Conference on Cognitive Informatics (pp. 255–260). IEEE. doi:10.1109/COGINF.2006.365705.

Ladyman, J., Lambert, J. and Wiesner, K. (2013). What is a complex system? *European Journal for Philosophy of Science* 3(1), 33–67. doi:10.1007/s13194-012-0056-8.

Lee, J. R. (2018). The fourth industrial revolution and future agriculture. In: Hussmann, S. (Ed.). *Automation in Agriculture* (p. 198). IntechOpen. doi:10.5772/intechopen.69016.

Letey, J., Hoffman, G. J., Hopmans, J. W., Grattan, S. R., Suarez, D., Corwin, D. L., Oster, J. D., Wu, L. and Amrhein, C. (2011). Evaluation of soil salinity leaching requirement guidelines. *Agricultural Water Management* 98(4), 502–506. doi:10.1016/j.agwat.2010.08.009.

Lu, Y., Liu, M., Li, C., Liu, X., Cao, C., Li, X. and Kan, Z. (2022). Precision fertilization and irrigation: progress and applications. *AgriEngineering* 4(3), 626–656. doi:10.3390/agriengineering4030041.

Lundstrom, D. and Stegman, E. (1995). *Irrigation Scheduling by the Checkbook Method*. North Dakota State University. NDSU Extension Service. Fargo, ND.

Maas, E. and Hoffman, G. (1977). Crop salt tolerance–current assessment. *Journal of the Irrigation and Drainage Division - ASCE*, 115–134. doi:10.1016/j.agwat.2010.08.009.

Marcelis, L., Costa, M. and Heuvelink, E. (2019). *Achieving Sustainable Greenhouse Production: Present Status, Recent Advances and Future Developments*. Burleigh Dodds Science Publisher. doi:10.19103/as.2019.0052.01.

Maroufpoor, S., Bozorg-Haddad, O. and Maroufpoor, E. (2020). Reference evapotranspiration estimating based on optimal input combination and hybrid artificial intelligent model: hybridization of artificial neural network with grey wolf optimizer algorithm. *Journal of Hydrology* 588, 125060. doi:10.1016/j.jhydrol.2020.125060.

Mmolawa, K. and Or, D. (2000). Root zone solute dynamics under drip. *Plant and Soil* 222(1/2), 163–190. doi:10.1023/A:1004756832038.

Mohan, S. and Arumugam, N. (1997). Expert system applications in irrigation management: an overview. *Computers and Electronics in Agriculture* 17(3), 263–280. doi:10.1016/S0168-1699(97)01309-4.

Morgan, K. (n.d.). *Citrus Microirrigation Scheduler*: Available at: https://fawn.ifas.ufl.edu/. Extension: University of Florida Retrieved 2022, from Florida Automated Weather Network.

Mundo, M., Mirelez, V., Martinez, P. and Zazueta, F. (1997). Sistema computacional para la distribución eficiente del agua en distritos de riego, SICODE v2.1. *Ingenieria Hidraulica en Mexico* 12(2), 29–36.

Muñoz-Carpena, R., Dukes, M. D., Li, Y. and Klassen, W. (2008). Design and field evaluation of a new controller for soil-water based irrigation. *Applied Engineering in Agriculture* 24(2), 183–191. Retrieved from https://elibrary.asabe.org/abstract.asp?aid=24266 (accessed March 9, 2023).

Naadimuthu, G., Raju, K. S. and Lee, E. S. (1999). A heuristic dynamic optimization algorithm for irrigation scheduling. *Mathematical and Computer Modelling*, 169–183. doi:10.1016/S0895-7177(99)00172-7.

Nassar, I. N. and Horton, R. (1992). Simultaneous transfer of heat, water, and solute in porous media: I. Theoretical development. *Soil Science Society of America Journal*, 1350–1356. doi:10.2136/sssaj1992.03615995005600050004x.

Nellingham, B. (2009). Method for irrigation scheduling based on soil moisture data acquisition. In: Knel, S. and Anderson, S. (Eds). *Irrigation District Sustainability* (pp. 383–400). Reno: U.S. Committee on Irrigation and Drainage.

Or, D. (1996). Drip irrigation in heterogeneous soils: steady state field experiments for stochastic model evaluation. *Soil Science Society of America Journal* 60(5), 1339–1349. doi:10.2136/sssaj1996.03615995006000050008x.

Parkash, V. and Singh, S. (2020). A review on potential plant-based water stress indicators for vegetable crops. *Sustainability* 12(10), 28. doi:10.3390/su12103945.

Parsons, L. and Morgan, K. (2022). Florida citrus microirrigation scheduler. Retrieved from Florida Automated Weather Network: https://fawn.ifas.ufl.edu/tools/irrigation/citrus/tables/ (accessed March 9, 2023).

Parsons, L. and Wheaton, A. (2000). Irrigation management and citrus tree response in a humid climate. *HortScience*35(6), 1043–1045.

Patel, J., Patel, H. and Bhatt, C. (2012). Fuzzy logic-based decision support system framework for irrigation scheduling. Nirma University International Conference on Engineering. Ahmedabad: IEEE. doi:10.1109/NUICONE.2012.6493286.

Pereira, L. S., Paredes, P. and Jovanovic, N. (2020). Soil water balance models for determining crop water and irrigation requirements and irrigation scheduling focusing on the FAO56 method and the dual Kc approach. *Agricultural Water Management* 241, 106357. doi:10.1016/j.agwat.2020.106357.

Perry, K. B. (1998). Basics of frost and freeze protection for horticultural crops. *HortTechnology* 8(1), 10–15. doi:10.21273/HORTTECH.8.1.10.

Peters, A., Iden, S. C. and Durner, W. (2019). Local solute sinks and sources cause erroneous dispersion fluxes in transport simulations with the convection–dispersion equation. *Vadose Zone Journal* 18(1). doi:10.2136/vzj2019.06.0064.

Phene, C. J. and Beale, O. W. (1976). High-frequency irrigation for water nutrient management in humid regions. *Soil Science Society of America Journal*, 430–436. doi:10.2136/sssaj1976.03615995004000030034x.

Philip, J. and de Vries, D. (1957). Moisture movement in porous materials under temperature gradients. *Transactions of the American Geophysical Union* 38(2), 222-232. doi:10.1029/TR038i002p00222.

Power, D. and Heavin, C. (2017). Decision Support, Analytics, and Business Intelligence. *Faculty Book Gallery* 178. Business Experts Press, New York. https://scholarworks.uni.edu/facbook/450.

Ravi, V., Williams, J. and Burden, J. (1998). Estimation of infiltration rate in the vadose zone: compilation of simple mathematical models. Report EPA/600/R-97/128a, US Environmental Protection Agency, National Risk Management Research Laboratory, Ada, OK.

Rawlins, S. L. (1973). Principles of managing high frequency irrigation. *Soil Science Society of America Journal* 37(4), 626–629. doi:10.2136/sssaj1973.03615995003700040041x.

Richards, L. A. (1931). Capillary conduction of liquids through porous medium. *Physics*, 318–333. doi:10.1063/1.1745010.

Rodrigues, K. V., Lima, L. A. and Thebaldi, M. S. (2020). Effects of fertigation on corrosion in galvanized steel used in center pivot systems. *Water Supply* 20(4), 1189–1194. doi:10.2166/ws.2020.029.

Rose, D. C., Sutherland, W. J., Parker, C., Lobley, M., Winter, M., Morris, C., Twining, S., Ffoulkes, C., Amano, T. and Dicks, L. V. (2016). Decision support tools for agriculture: towards effective design and delivery. *Agricultural Systems* 149, 165–174. ISSN 0308-521X. doi:10.1016/j.agsy.2016.09.009.

Saggi, M. K. and Jain, S. (2019). Reference evapotranspiration estimation and modeling of the Punjab northern India using deep learning. *Computers and Electronics in Agriculture* 156, 387–398. doi:10.1016/j.compag.2018.11.031.

Saggi, M. K. and Jain, S. (2022). A survey towards decision support system on smart Irrigation Scheduling using machine learning approaches. *Archives of Computational Methods in Engineering*. doi:10.1007/s11831-022-09746-3.

Saggi, M., Jain, S., Bhatia, A. and Sharda, R. (2022). Proposition of new ensemble data-intelligence model for evapotranspiration process simulation. *Journal of Ambient Intelligence and Humanized Computing* 24. doi:10.1007/s12652-021-03636-5.

Saravanan, T., Jha, S., Sabharwal, G. and Narayan, S. (2020). Comparative analysis of software life cycle models. 2020 2nd International Conference on Advances in Computing, Communication Control. and Networking (ICACCCN) (pp. 906–909). IEEE. doi:10.1109/ICACCCN51052.2020.9362931.

Schrerer, T. (n.d.). Web irrigation scheduler. Retrieved July 2022, from North Dakota Agricultural Weather Station: https://www.ag.ndsu.edu/waterquality/documents/web-based-irrigation-scheduler

Schwarzman, N. and Zur, B. (1986). Emitter spacing and geometry of wetted soil volume. *Journal of Irrigation and Drainage Engineering, ASCE* 11(2), 242–253. doi:10.1061/(ASCE)0733-9437(1986)112:3(242).

Sepúlveda, E., Zazueta, F., Xin, J. and Vergot, P. (2005). Developing extension education software for handheld computers. Proceedings of the Joint ASAE/CSAE (pp. 17-20). Tampa, FL: American Society of Agricultural Engineers. Retrieved from https://elibrary.asabe.org/abstract.asp?aid=19094 (accessed February 14, 2022).

Shamshiri, R., Kalantari, F., Ting, K., Thorp, K., Hameed, I., Weltzien, C., Ahmad, D. and Shad, Z. (2018). Advances in greenhouse automation and controlled environment agriculture: a transition to plant factories and urban agriculture. *International*

Journal of Agricultural and Biological Engineering 11(1), 1-22. doi:10.25165/j.ijabe.20181101.3210.

Sharma, G., Singh, A. and Jain, S. (2021). A hybrid deep neural network approach to estimate reference evapotranspiration using limited climate data. *Neural Computing and Applications* 34(5), 4013–4032. doi:10.1007/s00521-021-06661-9.

Singh, A. J. (2022). Smart technologies in livestock farming. In: Sehgal, S. S. (Ed.). *Smart and Sustainable Food Technologies* (pp. 22–27). Springer. doi:10.1007/978-981-19-1746-2_2.

Smajstrla, A. and Koo, R. (1986). Use of tensiometers for scheduling of citrus trickle irrigation. *Proceedings of the Florida State Horticultural Society* 99, 51–56. Retrieved from https://journals.flvc.org/fshs/article/download/94738/90761(accessed March 14, 2022).

Sørensen, C. G., Fountas, S., Nashf, E., Pesonend, L., Bochtis, D., Pedersen, S. M., Basso, B. and Blackmore, S. B. (2010). Conceptual model of a future farm management information system. *Computers and Electronics in Agriculture* 72(1), 37–47. doi:10.1016/j.compag.2010.02.003.

Strain, J. (1982). *Updated Inventory of Agricultural Computer Programs Available for Extension Use*. Gainesville, FL: Food and Resource Economics Department, Cooperative Extension Service, Institute of Food and Agricultural Sciences, University of Florida.

Strelkoff, T., Clemmens, A. and Schmidt, B. (2000). ARS software for simulation and design of surface irrigation. National Irrigation Symposium. Proceedings of the 4th Decennial Symposium (pp. 290–297). Phoenix, AZ: American Society of Agricultural Engineers. Retrieved from https://www.cabdirect.org/cabdirect/abstract/20003037400 (accessed February 5, 2023).

Subbaiah, R. (2013). A review of models for predicting soil water dynamics during trickle irrigation. *Irrigation Science* 31(3), 225–258. doi:10.1007/s00271-011-0309-x.

Sun, Y., Wang, C., Chen, H. and Ruan, H. (2020). Response of plants to water stress: a meta-analysis. *Frontiers in Plant Science* 11, 978. doi:10.3389/fpls.2020.00978.

Valdez-Aguilar, L. A. and Reed, D. W. (2007). Response of selected greenhouse ornamental plants to alkalinity in irrigation water. *Journal of Plant Nutrition* 30(3), 441–452. doi:10.1080/01904160601171983.

Weber, M. T., Pease, J. W., Vincent, W. H., Crawford, E. W. and Stilwell, T. C. (1983). *Microcomputers and Programmable Calculators for Agricultural Research in Developing Countries*. Michigan State University.

Weren, D. (1978). *Irrigation Scheduling: Checkbook Method*. Agricultural Extension Service. University of Minnesota. Minneapolis, MN.

White, R. E., Dyson, J. S., Haigh, R. A., Jury, W. A. and Sposito, G. (1986). A transfer function model of solute transport through soil. 2. Illustrative applications. *Water Resources Research* 22(2), 248–254.

World Economic Forum (2022). Fourth industrial revolution. Retrieved from Strategic Intelligence: https://intelligence.weforum.org/topics/a1Gb0000001RIhBEAW (accessed August 12, 2023).

Wu, T., Zhang, W., Jiao, X., Guo, W. and Hamoud, Y. (2021). Evaluation of stacking and blending ensemble learning methods for estimating daily reference evapotranspiration. *Computers and Electronics in Agriculture* 184, 1–17. doi:10.1016/j.compag.2021.106039.

Xin, J. and Zazueta, F. (1995). Real-time expert system for citrus microirrigation management. Proceedings of the 5th International Irrigation Congress (pp. 787-791). Orlando, FL: American Society of Agricultural Engineers.

Xin, J. and Zazueta, F. (2016). Technology trends in ICT-towards data-driven, farmer-centered and knowledge-based hybrid cloud architectures for smart farming. *Agricultural Engineering International: CIGR Journal* 14(4), 275–279. Retrieved from https://cigrjournal.org/index.php/Ejounral/article/view/3937/2496 (accessed October 23, 2022).

Yaseen, Z. M., Al-Juboori, A. M., Beyaztas, U., Al-Ansari, N., Chau, K.-W., Qi, C., Ali, M., Salih, S. Q. and Shahid, S. (2020). Prediction of evaporation in arid and semi-arid regions: a comparative study using different machine learning models. *Prediction of evaporation in arid and semi-arid regions: a comparative study using different machine learning models* 14(1), 70–89. doi:10.1080/19942060.2019.1680576.

Yu, L. W., Shamshiri, R., Tao, S., Ren, Y., Zhang, Y. and Su, G. (2021a). Review of research progress on soil moisture sensor technology. *International Journal of Agricultural and Biological Engineering* 14(4), 32–42. doi:10.25165/j.ijabe.20211404.6404.

Zhai, Z., Martínez, J. F., Beltran, V. and Martínez, N. L. (2020). Decision support systems for agriculture 4.0: survey and challenges. *Computers and Electronics in Agriculture* 170, 105256. ISSN 0168-1699. doi:10.1016/j.compag.2020.105256.

Zazueta, F. (1983). *HYPCHL: Diagnosis and Treatment of Iron and Slime Clogging Problems*. Gainesville, FL: IFAS, University of Florida.

Zazueta, F. (1992). *Microirrigacion*. Guadalajara: ICFA Int.

Zazueta, F. S. (1993). New technologies in the management of microirrigation systems. *Acta Horticulturae*, 305–312. doi:10.17660/ActaHortic.1993.335.37.

Zazueta, F. and Lokers, R. (1991). PCCONT: a microcomputer-based scheduler for nursery irrigation systems. University of Florida Cooperative Extension Service Circular 1066, 10. Gainesville, FL.

Zazueta, F. and Smajstrla, A. (1989). *Water Management Utilities*. Fla. Agric. Engr. Ext. Gainesville, FL: University of Florida.

Zazueta, F. and Xin, J. (1994). *Soil Moisture Sensors*. ABE. Gainesville, FL: UF/IFAS Extension.

Zazueta, F., Smajstrla, A. and Clark, G. (1988). Normalized irrigation intervals for Florida. *Soil and Crop Science Society of Florida Proceedings* 48, 96–99. Retrieved from https://p2infohouse.org/ref/13/12102.pdf (accessed September 23, 2022).

Zazueta, F., Smajstrla, A. and Harrison, D. (1985). A simple numerical model for the prediction of soil-water flow from trickle sources. *Soil and Crop Science Society of Florida Proceedings* 44, 72–76.

Chapter 7

Advances in crop disease forecasting models

Nathaniel K. Newlands, Summerland Research and Development Centre, Science and Technology Branch, Agriculture and Agri-Food Canada, Canada

1 Introduction

Diseases are widespread and destructive, causing severe yield quantity and quality losses, increasing agricultural production costs, and cumulative human health and environmental risks and impacts. The Food and Agriculture Organization of the United Nations estimates between 20% and 40% of global crop production lost annually are lost to diseases and pests. Each year, plant diseases cost the global economy around $220 billion. Plant diseases are primarily caused by pathogen infections by viruses, bacteria, and fungi. A total of identified 137 pathogens and pests explain losses of 1–40% in the staple crops (i.e. wheat, maize (corn), soybeans, rice, and potatoes), providing roughly 50% of the calorie intake among the world's humans (Savary et al., 2019). Global yield loss estimates for these major crops are 21.5% for wheat, 30% for rice, 22.5% for corn/maize, 17.2% for potato, and 21.4% for soybean and are highest in areas where disease frequently emerges or re-emerges, namely food-deficit regions with fast-growing populations. Climate change is anticipated to exacerbate current losses, by introducing complexity and uncertainty on our ability to adequately inform and effectively protect crops and ensure human populations are food, water, and energy secure. The

http://dx.doi.org/10.19103/AS.2023.0132.13
Published by Burleigh Dodds Science Publishing Limited, 2023.

consequences of climate change in relation to disease incidence and severity, epidemic frequency, and its spatial distribution on both wild plant communities, agricultural crops, and livestock/animal disease dynamics, are highly uncertain. Making agricultural production systems more resilient and sustainable requires a robust understanding of how to manage multiple diseases (and pests) in an integrated and effective way. Automated, precision agriculture technology and Earth observations (EO) provide new opportunities to advance crop disease monitoring, modelling, and forecasting.

Current, key issues and challenges in crop disease forecasting (in no priority order) are as follows:

- Establishing multidisciplinary collaboration between plant pathology and other disciplines (e.g. precision agriculture, bioinformatics, data science, statistics, climatology/meteorology) to advance knowledge, research questions, and solutions related to major aspects such as disease surveillance and detection technologies, risk model/algorithms, climate change biotic and abiotic impacts, biocontrol, host resistance, and microbial interaction.
- Long-term, proactive, and precautionary biovigilance strategies that involve disease surveillance to collect data across agricultural landscapes need to be designed and implemented to validate disease risk models and support operational, disease risk early warning systems (EWS).
- Ecological disease management needs to target diverse production situations and therefore must consider variation in attainable yields; this can be achieved through the reengineering of agroecosystems with higher biodiversity.
- A better understanding of how forecast uncertainty changes with spatial and temporal scale of information, such as data provided by climate service providers (i.e. observational, station-based data, gridded or interpolated station data, short-term meteorological model forecasts, coupled data-model assimilated 'reanalysis' products, and downscaled regional climate model data).
- The further exploration of artificial intelligence (AI)-based methods (i.e. machine learning (ML) and deep learning (DL)) for harnessing large, complex datasets for training and validating forecasts, reducing forecast error, and extending the accuracy and time windows of forecasts.
- More exploration of the impacts and consequences of human and disease behavioural dynamics and the broader assessment of forecast model accuracy and quality are needed. Agent-based, individual-based, or game-theoretic approaches offer potential to explain collective behaviour and observed, emergent multi-disease patterns explainable or to behavioural assumptions, dynamical rules, and real-world scenarios.

- Further research on plant-water-disease relationships as climate change progresses and ecological disease management needs to target diverse production situations, considering field, regional, and global variation in attainable yields under coupled climate-disease-crop risk scenarios.

2 Modelling complex, crop-disease-environment dynamics

Crop-disease-environment interactions are complex, adaptive, and dynamic, often involving multiple species and genotypes. These interactions exhibit behaviour that is not captured in simple infection models involving single host-single disease systems. Diseases cause monocyclic (single) and polycyclic (multiple) and mixed modal polycyclic infection cycles epidemics. Mixed-mode epidemics typically involve multiple hosts and vectors. Both endemic and emerging diseases are transported over long distances in the air and by human transmission and spill over within global food trade and transportation networks. For example, the brown streak viral disease of cassava, a major staple crop in Africa, exhibits mixed-mode transmission due to infectious material being transported by trade or sharing between farms in Uganda (McQuaid et al., 2017). The 'disease triangle' of any plant-pathogen-environment interaction may shift in response to climate change, favouring different climatic preferences and niche breadths in the future (Burdon and Zhan, 2020; Hardwick, 2006). Area residency and transboundary movement patterns are anticipated to change, leading to emerging and re-emerging outbreaks and upsurges, as new pathogens continuously evolve and adapt to changing environmental conditions. The current lack of an adequate understanding of how the mechanisms and interactions of multiple diseases cooperate, compete, and adapt is a major barrier to understanding changing their geographic patterns and forecasting climate change impacts (Burdon and Zhan, 2020). As described in detail by Abdullah et al. (2017), plant infections often involve multiple species and/or genotypes and exhibit complexities not captured in single host-single disease systems, yet studies of plant-pathogen interactions have historically focused on simple models of infection involving single host-single disease systems. Pathogen interactions include: (1) competition, in which competing pathogens develop physical barriers or utilize toxins to exclude competitors from resource-dense niches; (2) cooperation, whereby pathogens beneficially interact, by providing mutual biochemical signals essential for pathogenesis or through functional complementation via the exchange of resources necessary for survival; and (3) coexistence, whereby pathogens can stably coexist through niche specialization. Furthermore, crops, as hosts for multiple pathogens at a time, are also able to, actively or passively, modulate niche competition through defence responses that target at least one pathogen.

Forecasting is the practice of predicting future events or trends, considering present and past information, to inform decision-making, to gauge uncertainty and risk, while informing options, choices, actions, involved in reaching a defined goal. 'Prediction' is an actual act of indicating that something will happen in the future with or without prior information. All forecasts are predictions, but the opposite is not true as anyone can predict, but to forecast, analytical/modelling, reasoning, and interpretation skills are required. Disease forecasting involves predicting disease occurrence in a specified area ahead of time, allowing time to take appropriate disease control measures before losses may occur, and to guide management decisions to reduce disease risk, avoid initial inoculum, and/or slow down infection rates. The need for disease risk forecasting is increasingly important, as both endemic and emerging diseases are spread through transmission and spill over within global food trade networks. Forecasts are driven by historical and/or current trends and conditions, while projections are driven by scenarios (see Juroszek and von Tiedemann, 2015).

A general depiction of a forecast is shown in Fig. 1a. A training time window in which historical data are used to train a model to forecast in the future into a forecast time window is indicated. The number of observations is split into three separate datasets for model training, validation, and testing depending on signal-to-noise ratio of the data, the training sample size, and other data quality considerations. An example split typically allocates more data for model training, e.g. 50% for training, 25% validation, and 25% for testing. In model training, observational data are used to generate model predictions with a degree of prediction errors. A set (or ensemble) of model forecasts is generated by changing model assumptions, data inputs, and data-model calibration settings. Forecast error changes with model complexity (i.e. degrees of freedom), so under fitting and over fitting of models needs to be avoided to obtain the highest possible forecast accuracy, with minimal variance and bias in relation to reducible error, i.e. training a model to data (training error) and testing it to data (generalization error) (Fig. 1b). The generalization performance of a forecast model relates to its prediction capability on independent test data. Test error, also referred to as generalization error, is the prediction error over an independent test sample. 'Model selection' involves estimating the performance of different models in order to choose the best one, while 'model assessment', involves choosing a final model, estimating its prediction error on new data (Hastie et al., 2009). Plausible or counterfactual scenarios relating to one or many variables in a model and/or auxiliary indices from other models can be coupled into forecasts, to generate future or potential forecasted outcomes. For example, Cook et al. (2016) explored spatially explicit scenarios of different response options to Fire Blight within apple and pear orchards in the Goulburn Valley region of Victoria, Australia. This reduced costs and

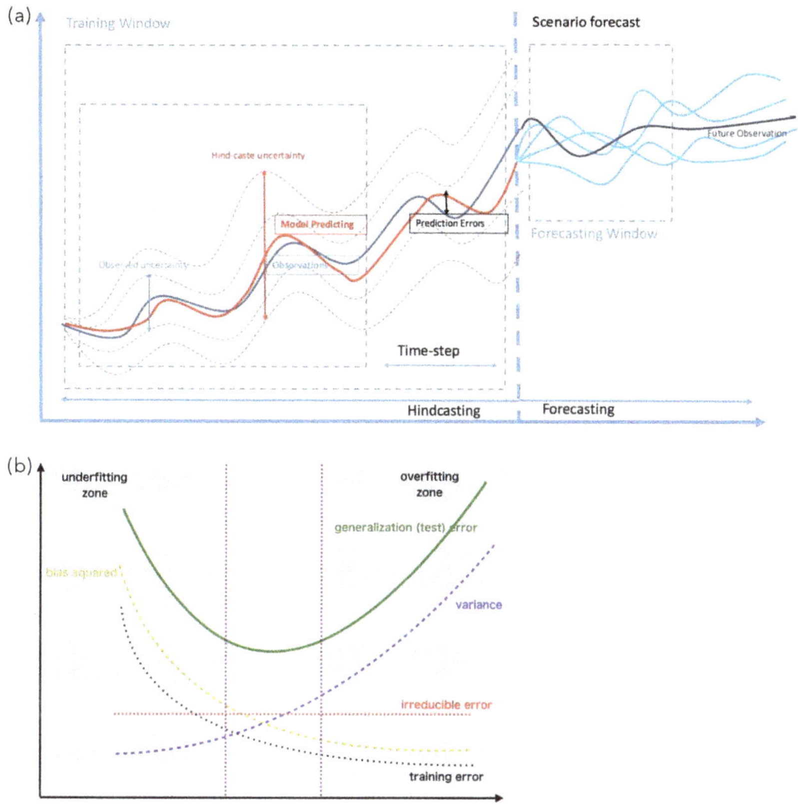

Figure 1 (a) Forecast output versus time. Forecast quality verification - attributes and diagnostics (Source: 'Representation of main aspects of statistical modelling predictions', Figure 1.1, page 3, Lu (2020), with permission by author). (b) Forecast error versus model complexity (degrees of freedom). Source: Papachristoudis, 2019.

enabled developing mutually satisfactory disease quarantine and destruction zones, along with consensus on the timing of withdrawing limited resources from eradication efforts.

Identifying the best model - one with the essential level of detail, complexity, and number of essential variables - for attaining a given target level of required forecast accuracy, remains a major challenge. There is no universal solution to this problem; instead relevant spatial, temporal scale and problem context are crucial. Overly simple (e.g. empirical or deterministic) models comprising a single driver or predictor variable may be inadequate because drivers change in their intensity and new drivers emerge. Statistical models may try to explain the dynamics of such changes in predictors but may not attain significant improvement in forecast accuracy. Furthermore, more complex, dynamic, or scenario (i.e. storyline) models that consider a broader set of potential drivers may offer projections but lack sufficient detail, data, and statistical support (Oliver and Roy, 2015). Recent work by Shah et al. (2019) involving wheat Fusarium head blight applies a scalar-on-function regression approach to address major statistical issues and problems with the traditional analysis of weather (and climate) time-series and associated changes with disease outcomes (termed window-pane, or time-window analysis involving exhaustive search) that can affect forecast accuracy and quality because of highly correlated predictors, spurious large correlations, and extreme multiplicity in training and testing of correlation. Skelsey (2021) has compared five different algorithms for detecting crop disease outbreaks in relation to weather conditions using historical data of potato light blight (PLB) (2012–17) for model training and hindcast testing. A mixture-model (Gaussian type) attained the highest forecast accuracy (97%) with added improvement when combing all algorithms, i.e. ensemble model forecasting. An overview of different methodologies for ecological (e.g. crop disease) forecasting is shown in Table 1. Models or model components may be integrative (e.g. ensemble-based), a mixture of models (i.e. mixture-models), or hybrid (e.g. AI-based and non-AI learning algorithms, empirical and probabilistic). A further discussion of AI algorithms and models is provided in Section 4.

Forecast quality has many defining attributes, linked with model calibration, refinement, resolution, discrimination, unconditional bias, and absolute and relative accuracy (see Murphy and Winkler, 1992). Verifying and validating a model forecast so that it is robust and reliable requires evaluating multiple performance measures, beyond just absolute forecast accuracy and relative forecast accuracy (or skill).

A conceptual framework for data-model fusion that addresses fundamental challenges related to human infectious disease, and also applicable or transferable to advancing plant or crop disease forecasting, proposes a hierarchical state-space structure, whereby: (1) multiple data sources and spatial scales are integrated to inform hidden or latent model parameters; (2) data and model uncertainty is partitioned by separate process and observation models; and (3) support is provided by scientific evidence from ecological and epidemiological knowledge (Ladeau et al., 2011). Forecasting the potential impacts of crop disease risk requires forecasting crop yield, as disease (pathogen), crop (host), and environment are intertwined and interdependent. An integrated framework for modelling and forecasting disease risk in relation to climate, extreme weather, crop disease infection cycling, and spray management control at the regional scale has been proposed (Newlands, 2018). This framework includes risk components for the major disease progression stages (i.e. immigration, deposition, germination, infection, incubation/residency, multiplication, re-dispersal, and perennation), alongside major risk variables, e.g. weather trigger events and conditions,

Table 1 Different methodologies for ecological (e.g. crop disease) forecasting. Models or model components may be integrative (ensemble) or hybrid, comprising a mix of methodologies.

Forecasting methodologies	Description	Example
Extrapolation	Descriptive statistical model of trend in system state variable	Predicting species geographic range margins from past rates of change
Statistical models	Statistical relationship between driver or pressure and intermediate processes that combine to determine the state	Predicting species distribution from relationships between occurrence and climate variables
Mechanistic models	Relationship between driver or pressure and intermediate demographic processes based on prior biological understanding	Predicting species abundance under climate changes based on physiological relationships between development rates and lethal temperatures
Agent- or individual-based models	Behavioural rules used to model individual decisions, often in combination with phenomological components relating to demography	Predicting impacts of climate change on individual movements and how this scales up to species range margin shifts
Artificial intelligence-based models	Machine- or deep-learning models	Disease detection from imagery data, disease classification linked with weather and climate conditions, disease infection dynamics, forecasting impacts of climate change on multi-disease risk, integrating big data into forecast models and early warning systems, collected from *in situ* field disease occurrence data, imagery data from unmanned aerial vehicles (drones), rovers/robots, and earth observational platforms

Source: Adapted from Oliver and Roy (2015), Table 1, page 774.

host exposure and susceptibility, pathogen survival, residency, latency, endemic infection, cycling, epidemic transmission and dispersal, and evolved virulence associated with the risk of resistance breakdown. Fernandes et al. (2019) present a model coupling method that enables combining several computational models within risk modelling frameworks to support integrated development, testing, and validation of pest and disease risk models. Using this approach, they have developed and tested the integration of a dynamic pest and disease damage model, coupling a wheat crop model (CSM) called CSM-NWheat, to the PEST module within the Decision Support System for

Agrotechnology Transfer (DSSAT) (https://dssat.net/). This enables accurately estimating and predicting of the potential impact of pests and diseases on wheat growth and development, and ultimately economic yield (Ferreira et al., 2019). It also extends the potential application of agroecosystem models, such as DSSAT, to address pest and disease risks for real-world scenario decision-making (Fig. 2).

Recent reviews on the use of models for exploring current and future impacts of pests and diseases on agricultural systems propose an integrated, multidisciplinary research roadmap to improve modelling and forecasting of the impacts caused by plant diseases and pests (Jeger et al., 2021; Donatelli et al., 2017). Major knowledge and research gaps need to be addressed to (1) improve the quality and availability of data for model inputs; (2) improve the quality and availability of data for model evaluation; (3) improve the integration with crop models; (4) improve the processes for model evaluation; and (5) develop a community of plant pest and disease modellers.

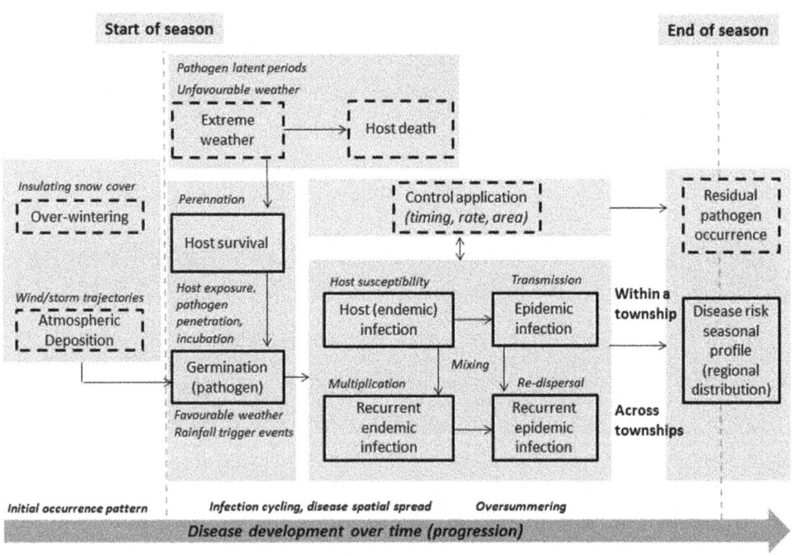

Figure 2 An integrated framework for modelling weather, disease infection cycling, and spray management control in modelling disease risk at the regional scale. Risk components for the major disease progression stages (i.e. immigration, deposition, germination, infection, incubation/residency, multiplication, re-dispersal, and perennation) are included, alongside major risk variables, e.g. weather trigger events and conditions, host exposure and susceptibility, pathogen survival, residency, latency, endemic infection, cycling, epidemic transmission and dispersal, and evolved virulence associated with the risk of resistance breakdown. Source: Newlands, 2018, Figure 1, page 4, Creative Commons (CC) licensing.

3 Big data assimilation to improve forecast quality

Big data (BD) refers to data that is machine-readable as opposed to human-readable, consisting of massive amounts of digital information, collected from all sorts of sources that are too large, raw, or unstructured for analysis using conventional relational database and techniques. There is no official size that makes data 'big'; however, BD needs to be of sufficiently high quality to reliably train, validate, and independently test and/or reproduce algorithmic and model output at reported levels of accuracy and reliability (Newlands, 2021a). For further details on governance, technology, and societal service delivery related to BD, see Newlands (2021b). Disease monitoring across large areas is typically costly and labour-intensive, relying on ground-based field scouting and visual identification by observers. Quinn et al. (2011) describe three general methods for improving the use of survey resources by performing data collection with mobile devices and by directing survey progress through the application of AI techniques in modelling and monitoring crop disease in developing countries. Precision agriculture using satellites, unmanned aerial vehicles, robots, spore detectors, and automated weather stations, combined with other sources of data, can improve the accuracy, efficiency, and effectiveness of disease monitoring, but requires a cost-effective, robust multi-resolution sampling design, reliable models, and agile operational systems for assimilating, analysing, and interpreting the data from these digital platforms and Internet-of-things (IoT) technologies. The IoT refers to the network of physical objects that exchange data between devices, software, and systems over the Internet. Ristaino et al. (2021) have outlined an integrated research agenda for mitigating future plant disease pandemics and identified critical components and data analytics needed for an emerging plant disease surveillance network that includes transportation and trade networks, geography, weather and climate parameters, and early detection data such as DNA sequence data, pathogen detection from sensors, text mining of historical and social media data, citizen science data, and identification applications. A better understanding of how to forecast uncertainty changes with spatial and temporal scale of information, such as data provided by climate service providers. Sources of weather and climate data include observational, station-based data, gridded or interpolated station data, short-term meteorological model forecasts, coupled data-model assimilated 're-analysis' products, and downscaled regional climate model data.

4 Novel artificial intelligence-based methodologies

AI refers to the simulation of human intelligence in machines that are programmed to think, learn, and problem-solve like humans and mimic their actions. ML is a subset of AI where machines learn from data without being explicitly programmed, and DL is a subset of ML in which artificial neural networks (ANNs) mimic the structure of the human brain to adapt and learn from vast amounts of data (Newlands, 2021a). Data science is an emerging domain of scientific inquiry, closely linked with applied mathematics, statistics, and AI that involve learning from data but is multidisciplinary and more computationally oriented, real-world task-driven, complex problem-oriented, and broader value-chain aware. The standard scientific inquiry cycle involves: (1) problem and hypothesis definition; (2) sampling plan/model identification; (3) data collection/processing; (4) analysis/modelling; and (5) interpretation/conclusions. Data scientists typically devote more time in the standard scientific inquiry cycle to the inclusion of modern technologies and BD, communicating uncertainty, actionable knowledge, and sharing insights across scientific disciplines and real-world application domains. An overview of a data science workflow for operational, AI-based disease risk forecasting and EWS decision-support is shown in Fig. 3.

The exploration of AI-based approaches for forecasting disease risk is in its infancy, with a need for developing new roadmaps for multidisciplinary investigation of AI forecasting and decision-support applications. In particular, there is a need for more exploration of the impacts and consequences of human and disease behavioural dynamics (Eksin et al., 2019), and a broader assessment of forecast model accuracy and quality is needed. Agent-based, individual-based, or game-theoretic approaches offer the potential to explain collective behaviour and observed, emergent multi-disease patterns explainable or to behavioural assumptions, dynamical rules, and real-world scenarios.

Fenu and Malloci (2019, 2021) have conducted a meta-analysis and classification of plant and crop forecasting models over the past 10 years (46 works, 2010-20). They found that the majority of these have focused on few pathogens and crops for predicting disease occurrence. A lack of required heterogeneous data across multiple diseases and crops, inter-comparison of AI and non-AI models, and consensus on real-world disease risk scenarios are hindering the advancement and scientific progress on crop disease forecasting.

Several recent studies exploring AI in disease modelling and forecasting are highlighted here:

- Pallagani et al. (2019) developed and tested a deep convolutional neural network model for classifying imagery of healthy and diseased plant leaves. They trained their models on a public dataset of 54,306 images, achieving a model training accuracy of 99% in identifying 26 diseases for 14 different crops. Such AI-based detection algorithms and models can provide automated detection when scouting or surveilling for disease, as

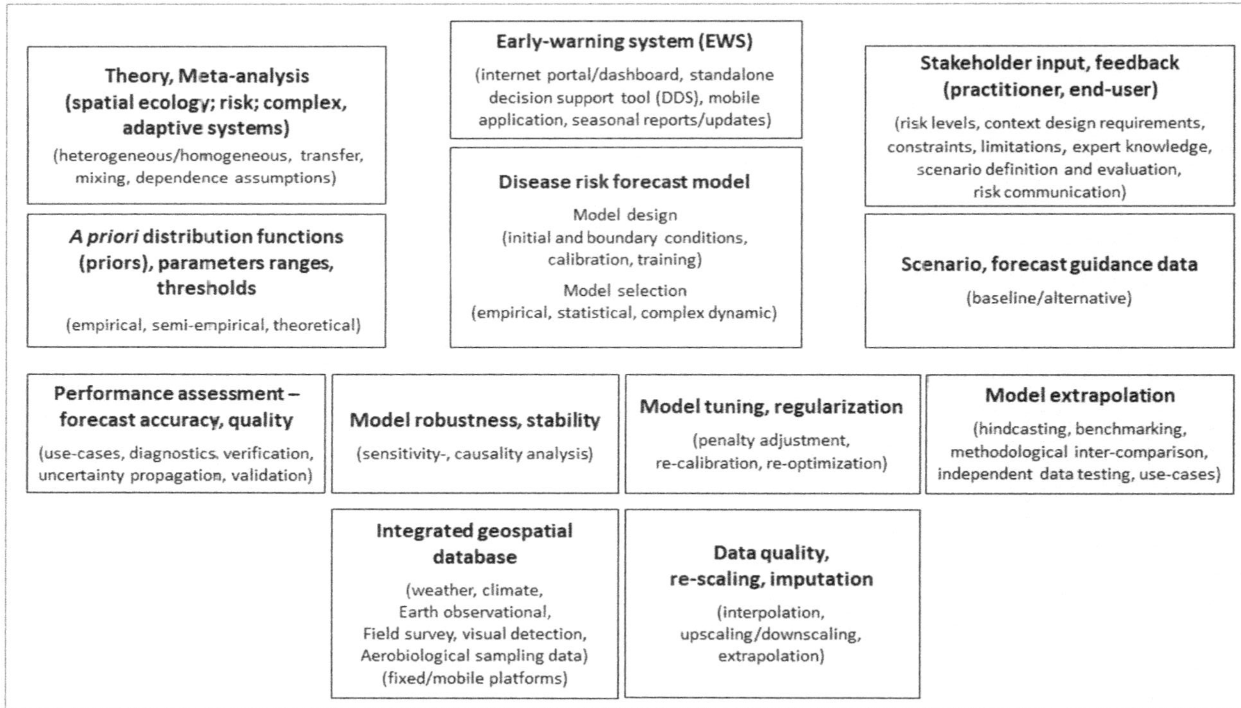

Figure 3 Overview of data science workflow for operational, artificial intelligence-based disease risk forecasting and early warning system decision-support. Source: original by author.

required to validate both model hindcasts (historical) and forecast (future) 'predictive' models of disease risk and occurrence.

- Fenu and Malloci (2019) reported that an ANN and Support Vector Machine (SVM) ML method predicting disease severity using 3 years of data for Sardinia improved the classification accuracy of a disease forecast model for PLB disease within 96–98%.
- Tilva et al. (2013) conducted a pilot study, applying an expert knowledge- or rule-based system that uses 'fuzzy' logic for predicting the risk of multi-diseases of corn (e.g. rust, spot) based on favourable weather conditions (multi-variate data of temperature, humidity, leaf wetness duration). Further work on the use of fuzzy logic expert systems for improving disease risk forecasts is needed.
- Lu et al. (2020) developed and validated a probabilistic, Bayesian learning network model to explore the complex causal interactions between environment, pathogen, and host for three different susceptible northern grape cultivars in Quebec, Canada. The model combines environmental (weather, climate), pathogen (development stages), and host (crop cultivar-specific susceptibility) factors. The model is evaluated in an operational forecast mode with supervised and algorithm model learning and integrating Global Forecast System Ensemble Reforecasts.

5 Case study: operational, crop disease early warning systems

EWS are an adaptive measure for climate change, using integrated communication systems, to help communities prepare for hazardous climate-related events, supporting long-term sustainability. They can save lives, livelihoods, resources, and infrastructure. Enhancements of EWS and the inclusion of forecast models provide a way to increase stakeholder and broader societal awareness, strengthen emergency and disaster risk communication, warnings, and alerts, harnessing BD and both human and AI-based intelligence for improved complex (i.e. multi-scale, multi-agency, multi-variate) decision-making under uncertainty. Areas of current research include: (1) the use of geospatial models and risk maps; (2) developing and testing cost-effective monitoring systems; (3) data assimilations from sensors; (4) improve model predictive power; (5) decision support tool design, deployment, and community adoption; (6) communication methods; (7) human response behaviour; and (8) vulnerability, exposure, impact visualization, and assessment. The primary purpose of a disease risk EWS is to warn farmers that the impending onset of infection period has already occurred so that they can take immediate appropriate control measures to stop further infection (Murmu et al., 2021).

Published by Burleigh Dodds Science Publishing Limited, 2023.

The impact of disease in reducing crop yields by directly damaging harvestable plant material (e.g. measured by survey estimates of yield or harvest index) has received much less attention than reductions in radiation interception and radiation use efficiency (Savary and Willocquet, 2020). Several major global EWS are now operational providing actual crop yield reports and attainable crop yield forecasts. The Crop Monitor for Early Warning is an EWS deployed by the G20 Group on Earth Observations Global Agriculture Monitoring, an international initiative to strengthen agricultural, humanitarian intervention, and food security decision-making and policy implementations. The Nutrition Early Warning System (CGIAR Platform for BD in Agriculture) International Center for Tropical Agriculture aggregates and analyses satellite imagery and traditional data, such as rainfall, temperature, and vegetation health, to help predict the nutritive value of crops.

There are also many different 'disease' forecasting models and EWS tools now operational worldwide. Information about the spread of crop disease is vital for both developed and developing countries. Surveys are slow, expensive, and typically rely on scarce resources and labour to gather data generating insufficient quantities of high-quality data needed for training and validating predictive algorithms and models for forecasting crop yield, yield quality, and disease.

- The Wheat Blast Early Warning System developed by the Brazilian Agricultural Research Corporation and the International Maize and Wheat Improvement Center for Brazil and Bangladesh (Fernandes et al., 2021) (https://beattheblastnews.net).
- CropIn's Disease Early Warning System[1] is a weather-based disease prediction technology.
- A national-scale, geospatial disease risk forecasting system for China that integrates Earth Observational and other multisource data (Dong et al., 2020).
- United States Department of Agriculture EWS for Soybean rust in real-time at the county-scale (Roberts et al., 2006).
- An AI-based EWS for grape downy mildew that successfully reduced fungicide treatments by 50% in Bordeaux vineyards, compared to current agronomic practices (Chen et al., 2020).
- An example EWS that forecasts and mitigates wheat rust diseases in Ethiopia, integrates combines multidisciplinary expertise across multiple organizations, novel models and tools, and BD from field and mobile phone surveillance data, spore dispersal and disease environmental suitability, and existing environmental research infrastructures to provide operational forecasts and disease risk communications for policy-makers, advisors, and smallholder farmers (Allen-Sader et al., 2019) (Fig. 4).

1 https://www.cropin.com/blogs/cropins-disease-early-warning-system-dews.

6 Conclusion and future trends

A proactive, precautionary 'biovigilance' strategy integrates awareness, detection, assessment and prioritization, understanding, mitigation, and appropriateness for monitoring unintended effects and anticipating future disease risks threats (AAFC, 2022; Odile et al., 2017). On-going surveillance, operational model-based forecasting, and EWS are crucial components and tools to support a biovigilance disease and pest risk management strategy, alongside adaptive management practices involving biological control, clean technologies, and integrated pest management. In addition to the continuation of field-scale research experiments (e.g. randomized control trials), long-term automated monitoring networks and community-based surveillance networks, at the farm to regional scale, can provide timely, enhanced, and consistent data with appropriate coverage (in space and time) for improving the timeliness,

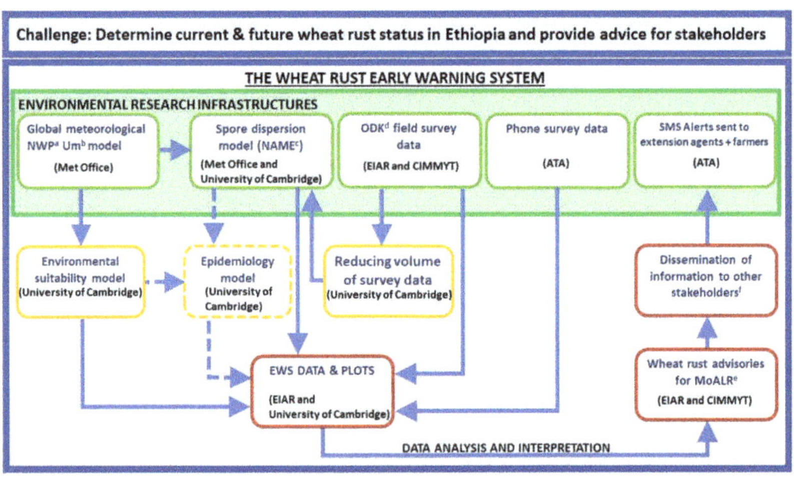

Figure 4 Overview of the Ethiopian wheat rust early warning system that integrates information from numerical weather forecasts, atmospheric-dispersion models, operational. Environmental research infrastructures from the different project partners and how together with other components, these make up the Ethiopian wheat rust early warning system. Solid lines indicate the initial structure of information flows and dashed box and arrows indicate phase 2 flows in the next version of the early warning system. Green outline boxes indicate existing environmental research infrastructures. Yellow outline boxes show the new tools and models developed (or future planned development) especially for this wheat rust early warning system. Red boxes indicate the early warning system output. Numerical Weather Prediction. Unified Model Numerical Atmospheric-dispersion Modelling Environment Open Data Kit. Ministry of Agriculture and Livestock Resources. Other stakeholders include Federal and Regional government agencies, research institutes, non-governmental organizations (NGOs), and Food and Agriculture Organization of the United Nations (FAO). Source: Clare Allen-Sader et al. (2019), Figure 1, page 3, Creative Commons (CC) licensing.

quality, usefulness, and relevance of model-based forecasts of disease risk and operational EWS. There is an urgent need to integrate multisource, digital agricultural BD and AI-based models in forecasting models in agriculture. While changing current planting times or the amounts and timing of pesticide applications may help to lessen disease impacts, the effectiveness of pesticides (e.g. fungicides) continues to decline as pathogens develop resistance (Odile et al., 2010). Designing genetically modified crops, such as new varieties that are climate-resistant to drought, heat, environmental conditions such as submersion and salinity, and pest-resistant, using genetic engineering (e.g. clustered regularly interspaced short palindromic repeat genome-editing technology) opens new opportunities to engineer disease resistance traits (Zaidi et al., 2020). However, this introduces further unknown risks and uncertainty, as the longer-term, eco-evolutionary consequences of climate change on pathogens of both GM and non-GM crops are not well understood. Developing, testing, and introducing novel resistant varieties involve long regulatory timelines, legal hurdles, high commercialization costs, and societal concerns and responses in regard to potentially accelerating pathogen adaptation and emergence of more harmful lineage, unknown and unanticipated environmental and human health impacts (see review by Zaidi et al. (2020) and references therein). Broader availability, learning, and awareness of digital agricultural data, predictive methods, and tools worldwide are taking place through greater connectivity, bandwidth, and computational speed. Advancing our knowledge and developing better, smarter tools for reliably and efficiently monitoring, forecasting, and detecting crop disease, provide a viable, risk-averse, and low-cost (and possibly less destructive) approach for anticipating, preventing, and mitigating crop disease infection, outbreaks, and pandemics.

7 Where to look for further information

7.1 Further reading

- Chaube, H. (2017). *Plant Disease Management: Principles and Practices* (1st ed.). Routledge, Taylor & Francis Group. 335pp. ISBN 9781315896625, doi: 10.1201/9781351075725.
- Dietze, M. C. (2017). *Ecological Forecasting*. Princeton: Princeton University Press, 270 pp. ISBN: 9780691160573.
- Fernandes, J. M. C., Pavan, W., Pequeno, D., Wiest, R., Holbig, C. A., Oliveira, F. and Hoogenboom, G. (2019). Improving crop pest/disease modeling, pp 1–23, In: *Advances in Crop Modelling for a Sustainable Agriculture*. (Boote, K, Ed.), Burleigh Dodds Series in Agricultural Science, Burleigh Dodds Science Publishing Ltd. doi: 10.19103/AS.2019.0061.07. Available

online: http://www.alice.cnptia.embrapa.br/alice/handle/doc/1119502 (Accessed May 10, 2022).

- Hastie, T., Tibshirani, R. and Friedman, J. H. (2009). *The Elements of Statistical Learning: Data Mining, Inference, and Prediction*. 2nd ed. New York: Springer, 764pp.
- Scott, P., Strange, R. Korsten, L. and Gullino, M. L. (2021). *Plant Diseases and Food Security in the 21st Century*. Vol. 10. 1st ed. Springer International.
- Singh K. P., Jahagirdar, S. and Sarma, B. K. (eds.) (2021). *Emerging Trends in Plant Pathology*. Singapore: Springer Nature, 844pp. doi: 10.1007/978-981-15-6275-4.

7.2 Key journals/conferences

- *Plant Disease and Phytopathology* (Official journal of The American Phytopathological Society, APS) https://apsjournals.apsnet.org/.
- *International Journal of Forecasting*, Elsevier (Official journal of International Institute of Forecasters) https://www.journals.elsevier.com/international-journal-of-forecasting.
- *Environmetrics*, Wiley (Official Journal of The International Environmetrics Society (TIES), an Association of the International Statistical Institute) https://onlinelibrary-wiley-com.ezproxy.library.uvic.ca/journal/1099095x.
- Agriculture and Climate Change Conference, Elsevier (4th Annual conference upcoming on 7-9 May 2023 in Dresden, Germany) https://www.elsevier.com/events/conferences/agriculture-and-climate-change-conference.
- Global Conference 2023, Elsevier (The International Society for Ecological Modelling) 'Ecological Models for Tomorrow's Solutions', 2-6 May 2023, University of Toronto, Scarborough, Canada https://www.elsevier.com/events/conferences/international-society-for-ecological-modelling-global-conference.

7.3 Major international research projects

- **Global Crop Pest and Disease Monitoring & Forecasting (PEST& DISEASE)** (2020-2022) is a community activity in the work programme of the Group on Earth Observations (GEO),[2] a partnership of more than 100 national governments and in excess of 100 participating organizations that envisions a future where decisions and actions for the benefit of humankind are informed by coordinated, comprehensive, and sustained Earth observations (EO). The project is a collaboration of the

2 GEO Work Programme (2020-2022): https://www.earthobservations.org/geoss_wp.php.

Institute of Remote Sensing and Digital Earth (RADI), Chinese Academy of Sciences, CAB International (CABI), Manchester Metropolitan University (MMU), and Ministry of Agriculture's National Agro-Tech Extension and Service Centre, China (NATESC). The project focuses on aphid and rust for wheat, planthopper and blast for rice, and aphid and mosaic virus for soybean. The project investigates the integration of multisource information, including EO, meteorological, biological, and plant protection datasets, for sustainable agricultural management decision-making support. It seeks to improve the utilization of EO data for forecasting pest and disease through the development of new algorithms and the fusion of new and existing data products of EO data. It will develop new algorithms, a novel global analytical framework, risk forecasting methodology, and geospatial risk maps of disease and pest outbreak forecasts.

- **Project summary:** https://www.earthobservations.org/documents/gwp20_22/CROP-PEST-MONITORING.pdf.
- **Global Burden of Crop Loss** (2019–2026) is a global project led by the Centre for Agriculture and Bioscience International (CABI or CAB International), an international, inter-governmental, not-for-profit organization with a global network of centres in 49 member countries. CABI is dedicated to improving people's lives worldwide by providing information and applying scientific expertise to solve problems in agriculture and the environment. This project, focusing on crop loss due to pest and disease, is funded by Grand Challenges programme, an initiative of the Bill & Melinda Gates Foundation, with the aim of generating comprehensive, authoritative evidence on the causes and risk factors of crop loss worldwide. This will allow agricultural research, policy, funding, and interventions, to maximize impact through improved targeting and relevance.
- **Project site:** http://www.croploss.org.
- **Project summary:** https://www.cabi.org/wp-content/uploads/Global-Burden-of-Crop-Loss-flyer-FINAL.pdf.

8 References

AAFC (2022). Vision statement. Biovigiliance: a modern approach to plant pest management. Agriculture and Agri-Food Canada (AAFC), National Biovigiliance Working Group, p. 29 Internal Report for Public Release in 2022-23.

Abdullah, A. S., Moffat, C. S., Lopez-Ruiz, F. J., Gibberd, M. R., Hamblin, J. and Zerihun, A. (2017). Host-multi-pathogen warfare: pathogen interactions in co-infected plants. *Front. Plant Sci.* 8: 1806. doi: 10.3389/fpls.2017.01806.

Allen-Sader, C., Thurston, W., Mayer, M., Nure, E., Bacha, N., Alemayehu, Y., Stutt, R. O. J. H., Safka, D., Craig, A. P., Derso, E., Burgin, L. E., Millington, S. C., Hort, M. C., Hodson,

D. P., Gilligan, C. A., Safka, D., Craig, A. P. and Derso, E. (2019) An early warning system to predict and mitigate wheat rust diseases in Ethiopia. *Environ. Res. Lett.* 14(11): 115004. doi: 10.1088/1748-9326/ab4034.

Burdon, J. J. and Zhan, J. (2020). Climate change and disease in plant communities. *PLoS Biol.* 18(11): e3000949. doi: 10.1371/journal.pbio.3000949.

Chen, M., Brun, F., Raynal, M. and Makowski, D. (2020). Forecasting severe grape downy mildew attacks using machine learning. *PLoS ONE* 15(3): e0230254. doi: 10.1371/journal.pone.0230254.

Cook, D. C., Aurambout, J., Villalta, O. N., Liu, S., Edwards, J. and Maharaj, S. (2016). A bioeconomic 'war game' model to simulate plant disease incursions and test response strategies at the landscape scale. *Food Sec.* 8(1): 37–48. doi: 10.1007/s12571-015-0524-z.

Donatelli, M., Magarey, R. D., Bregaglio, S., Willocquet, L., Whish, J. P. M. and Savary, S. (2017). Modelling the impacts of pests and diseases on agricultural systems. *Agric. Syst.* 155: 213–224. doi: 10.1016/j.agsy.2017.01.019.

Dong, Y., Liu, L., Du, X., Ren, B., Guo, A., Geng, Y., Ruan, C., Ye, H., Huang, W. and Zhu, Y. (2020). Automatic system for crop pest and disease dynamic monitoring and early forecasting. *IEEE J. Sel. Top. Appl. Obs. Remote Sens.* 13: 4410–4418.

Eksin, C., Paarporn, K. and Weitz, J. S. (2019). Systematic biases in disease forecasting – the role of behavior change. *Epidemics* 27: 96–105. doi: 10.1016/j.epidem.2019.02.004.

Fenu, G. and Malloci, F. M. (2019). An application of machine-learning technique in forecasting crop disease. In: *Proceedings of the 2019 3rd International Conference on Big Data Research (ICBDR 2019).* New York, NY: Association for Computing Machinery, pp. 76–82. doi: 10.1145/3372454.3372474.

Fenu, G. and Malloci, F. M. (2021). Forecasting plant and crop disease: an explorative study on current algorithms. *Big Data Cogn. Comput.* 5(1): 1–24. doi: 10.3390/bdcc5010002.

Fernandes, J. M., Del Ponte, E. M., Ascari, J. P., Krupnik, T. J., Pavan, W., Vargas, F. and Berton, T. (2021). Towards an early warning system for wheat blast: epidemiological basis and model development. In: *Burleigh Dodds Series in Agricultural Science*, pp. 623–642. doi: 10.19103/AS.2021.0092.22. Available at: https://bdspublishing.com/_webedit/uploaded-files/All%20Files/Open%20Access/9781801461733.pdf (accessed May 10, 2022).

Ferreira, T. B., Pavan, W., Fernandes, J. M. C., Asseng, S., Antunes de Oliveira, F., Hölbig, C. A., Pequeno, D. N. L. and Dalmago, G. A. (2019). Coupling a pest and disease damage module with CSM-N wheat: a wheat crop simulation model. *Trans. Am. Soc. Agric. Biol. Eng. (ASABE)* 64(6): 2061–2071. Available at: https://elibrary.asabe.org/abstract.asp?aid=52974 (accessed May 10, 2022).

Hardwick, N. V. (2006). Disease forecasting. In: Jones, B. M., Gareth, D. and Kaye, B. (Eds.), *The Epidemiology of Plant Diseases Cooke* (2nd edn.). The Netherlands: Springer, pp. 239–267.

Hastie, T., Tibshirani, R. and Friedman, J. H. (2009). *The Elements of Statistical Learning: Data Mining, Inference, and Prediction* (2nd edn.). New York: Springer, p. 764.

Jeger, M., Beresford, R., Bock, C., Brown, N., Fox, A., Newton, A., Vicent, A., Xu, X. and Yuen, J. (2021). Global challenges facing plant pathology: multidisciplinary approaches to meet the food security and environmental challenges in the mid-twenty-first century. *CABI Agric. Biosci.* 2(1): 20. doi: 10.1186/s43170-021-00042-x.

Juroszek, P. and von Tiedemann, A. (2015). Linking plant disease models to climate change scenarios to project future risks of crop diseases: a review. *J. Plant Dis. Prot.* 122(1): 3–15. Available at: http://www.jstor.org/stable/24618923.
Ladeau, S. L., Glass, G. E., Hobbs, N. T., Latimer, A. and Ostfeld, R. S. (2011). Data-model fusion to better understand emerging pathogens and improve infectious disease forecasting. *Ecol. Appl.* 21(5): 1443–1460. doi: 10.1890/09-1409.1.
Lu, W. (2020). *Multi-risk modeling for improved agriculture decision-support: predicting crop yield variability and gaps due to climate variability, extreme events, and disease.* Ph.D. thesis. University of Victoria, Canada. Available at: https://dspace.library.uvic.ca/bitstream/handle/1828/12130/Lu_Weixun_PhD_2020.pdf?sequence=1&isAllowed=y (accessed June 3, 2022).
Lu, W., Newlands, N. K., Carisse, O., Atkinson, D. E. and Cannon, A. J. (2020). Disease risk forecasting with Bayesian learning networks: application to grape powdery mildew (*Erysiphe necator*) in vineyards. *Agronomy* 10(5): 622. doi: 10.3390/agronomy10050622.
McQuaid, C. F., van den Bosch, F., Szyniszewska, A., Alicai, T., Pariyo, A., Chikoti, P. C. and Gilligan, C. A. (2017). Spatial dynamics and control of a crop pathogen with mixed-mode transmission. *PLoS Comput. Biol.* 13(7): e1005654. doi: 10.1371/journal.pcbi.1005654.
Murmu, R., John, V. and Maurya, A. K. (2021). Chapter 5: current research and innovations in plant pathology. In: Singh, H. K. (Ed.), *Plant Disease Forecasting System* (vol. 10). New Delhi: AkiNik Publications, pp. 71–84. doi: 10.22271/ed.book.794.
Murphy, A. H. and Winkler, R. L. (1992). Diagnostic verification of probability forecasts. *Int. J. Forecast.* 7(4): 435–455.
Newlands, N. K. (2018). Model-based forecasting of agricultural crop disease risk at the regional scale, integrating airborne inoculum, environmental, and satellite-based monitoring data. *Front. Environ. Sci.* 6: 63. doi: 10.3389/fenvs.2018.00063.
Newlands, N. K. (2021a). *Artificial Intelligence and Big Data Analytics in Vineyards - A Review.* Edited by Grapes, W., Morata, A., Loira, I. and Gonzalez, C. IntechOpen. ISBN: 978-1-83969-642-8. doi: 10.5772/intechopen.99862.
Newlands, N. K. (2021b). Big data governance, technology, and implementation in climate-resilient societies. In: Brears, R. C. (Eds.), *The Palgrave Handbook of Climate Resilient Societies.* Cham: Palgrave Macmillan, pp. 25–21. doi: 10.1007/978-3-030032811-5.
Odile, C., Fall, M. L. and Vincent, C. (2017). Using a biovigilance approach for pest and disease management in Quebec vineyards,. *Can. J. Plant. Pathol.*, 39(4):393--404, doi: 10.1080/07060661.2017.1366368.
Odile, O., Carisse, C., David-Mathieu, D., Tremblay, T., Tristan, T., Jobin, J., Sophie, A. and Walker, W. (2010). Disease decision support systems: their impact on disease management and durability of fungicide effectiveness. In: *Fungicides.* IntechOpen. doi: 10.5772/13335.
Oliver, T. H. and Roy, D. B. (2015). The pitfalls of ecological forecasting. *Biol. J. Linn. Soc. Lond.* 115(3): 767–778. doi: 10.1111/bij.12579.
Pallagani, V., Khandelwal, B. C., Udutalapally, V., Das, D. and Mohanty, S. P. (2019). *dCrop: A Deep-Learning Based Framework for Accurate Prediction of Diseases of Crops in Smart Agriculture, IEEE International Symposium on Smart Electronic Systems (iSES) (Formerly iNiS)*, pp. 29–33. doi: 10.1109/iSES47678.2019.00020.
Papachristoudis, G. (2019). The bias-variance tradeoff. Towards data science. Available at: https://towardsdatascience.com/the-bias-variance-tradeoff-8818f41e39e9 (accessed June 3, 2022).

Quinn, J. A., Leyton-Brown, K. and Mwebaze, E. (2011). Modeling and monitoring crop disease in developing countries. In: *AAAI Proceedings of the 25th AAAI Conference on Artificial Intelligence (AAAI)*. AAAI Press 25(1): 1390–1395. Available at: https://www.aaai.org/ocs/index.php/AAAI/AAAI11/paper/view/3777/4083 (accessed May 3, 2022).

Ristaino, J. B., Anderson, P. K., Bebber, D. P., Brauman, K. A., Cunniffe, N. J., Fedoroff, N. V., Finegold, C., Garrett, K. A., Gilligan, C. A., Jones, C. M., Martin, M. D., MacDonald, G. K., Neenan, P., Records, A., Schmale, D. G., Tateosian, L. and Wei, Q. (2021). The persistent threat of emerging plant disease pandemics to global food security. *Proc. Natl Acad. Sci. U. S. A.* 118(23): e2022239118. doi: 10.1073/pnas.2022239118.

Roberts, M. J., Schimmelpfennig, D., Ashley, E. and Livingston, M. (2006). *The Value of Plant Disease Early-Warning Systems*. Economic Research Report No. 18. Economic Research Service of the United States Department of Agriculture (USDA), p. 46. Available at: https://www.ers.usda.gov/publications/pub-details/?pubid=45314 (accessed May 3, 2022).

Savary, S. and Willocquet, L. (2020). Modeling the impact of crop diseases on global food security. *Annu. Rev. Phytopathol.* 58: 313–341. doi: 10.1146/annurev-phyto-010820-012856.

Savary, S., Willocquet, L., Pethybridge, S. J., Esker, P., McRoberts, N. and Nelson, A. (2019). The global burden of pathogens and pests on major food crops. *Nat. Ecol. Evol.* 3(3): 430–439. doi: 10.1038/s41559-018-0793-y.

Shah, D. A., Paul, P. A., De Wolf, E. D. and Madden, L. V. (2019). Predicting plant disease epidemics from functionally represented weather series. *Phil. Trans. R. Soc. B Biol. Sci.* 374(1775): 20180273. doi: 10.1098/rstb.2018.0273.

Skelsey, P. (2021). Forecasting risk of crop disease with anomaly detection algorithms. *Phytopathology* 111(2): 321–332. doi: 10.1094/PHYTO-05-20-0185-R.

Tilva, V., Patel, J. and Bhatt, C. (2013). Weather based plant diseases forecasting using fuzzy logic. In: *Nirma University International Conference on Engineering (NUiCONE)* (vol. 2013), pp. 1–5. doi: 10.1109/NUiCONE.2013.6780173.

Zaidi, S., Se, A., Mahas, A., Vanderschuren, H. and Mahfouz, M. M. (2020). Engineering crops of the future: CRISPR approaches to develop climate-resilient and disease-resistant plants. *Genome Biol.* 21: 289. doi: 10.1186/s13059-020-02204-y.

Chapter 8

Smart farming in extensive livestock production: the Australian experience

David W. Lamb, Food Agility Cooperative Research Centre/Precision Agriculture Research Group - University of New England/Gulbali Research Institute - Charles Sturt University, Australia

1 Introduction

Australia, with its principal land masses covering 7.69 million square km and lying between 11 degrees and 44 degrees south, is approximately 3800 km long (N-S) and 4000 km wide (E-W). It is considered the oldest, flattest and driest continent on earth. Climate zones include tropical, arid, temperate, cold and polar tundra (Beck et al., 2018). Agricultural production contributes AU$24.7 billion to Australia's GDP (AU$985 per capita) (ABARES, 2021, 2023). Dominant industries include red meat (cattle and sheep), arable crops and horticulture.

http://dx.doi.org/10.19103/AS.2023.0132.02

Approximately 55% of Australia's landmass is used for agricultural production; when taken with an estimated 87 800 farm businesses, this equates to an average farm size of 4863 ha. According to global census data available at the time of analysis (Lowder et al., 2016), Australia has the largest average farm size of any country; 5.5 times that of the next largest, Argentina, and 34 times that of the largest European country (Sweden). By virtue of the intensity of labour and other inputs per hectare for our comparatively large farming enterprises, the vast majority of our production comes from 'extensive farming' and it is for this reason that this chapter will focus on smart farms from the perspective of extensive farming.

Land use and land-use sustainability is a hot topic in Australia both for farm business and the public more broadly (Coelli, 2021). Climate change is a clear and present challenge. For example, Hughes et al. (2022) estimate that changes in seasonal conditions over the 20-year period up to 2020 have reduced annual average broadacre farm profits by 23%. As elsewhere in the world (Fuglie, 2018; Annosi et al., 2020), productivity growth in Australian agriculture has largely been led by improvements in technology and structural change (ABARES, 2021). Of course, productivity and efficiencies in agricultural production are never homogeneous across production units, namely farms. For example, in Australia, it has been estimated that lifting the output per hectare of all broadacre farms to that of the top 20% of producers would increase the total value of broadacre agricultural output by 24% above current levels, and farm cash income to 46% above current levels (ABARES, 2021).

The role of digital technology in raising per farm, and hence sector-wide production in Australia is substantive and the benefits substantial. In Australia, for example, the agriculture, fisheries and forestry sector aims to achieve AU$100 billion in annual farm gate output by 2030 (up from the present value of AU$61 billion) (DAWE, 2020). Full adoption of digital tools by the agriculture sector could contribute approximately $20.3 billion each year (Perrett et al., 2017). It is acknowledged, however, that this uplift requires work to improve the sectors' readiness to adopt new technologies, build a digitally capable and equipped agriculture workforce and improve the use of data for decision-making (Leonard et al., 2017; Cook et al., 2021).

Australia has enacted recent national and state policy changes aimed at increasing investment in digital innovation and driving adoption of digital solutions to support increasing productivity, for example through Agricultural Innovation Australia (AIA) and the 15 Research and Development Corporations (https://minister.awe.gov.au/littleproud/media-releases/cross-industry-agricultural-innovation) and the Victorian Government Digital Agriculture Strategy 2018 (https://agriculture.vic.gov.au/__data/assets/pdf_file/0006/567051/Digital-agriculture-strategy-2018.pdf). Data is a key area for improvement across the sector. The National Farmers Federation has developed their Farm Data Code (https://nff.org.au/programs/australian-farm-data-code/) as an early step toward providing more certainty around data exchange and access. The *Get Australia Growing* strategy is intended to help build trust in farmers to share data necessary to drive analytics behind many technology platforms (https://nff.org.au/getaustraliagrowing/) and the recently-launched *AgTrace* initiative will focus on permissions-based data sharing for facilitating export market access (https://www.foodagility.com/posts/agtrace-is-improving-traceability-for-australias-agriculture-sector). Underpinning data policies and infrastructure development is the need for improved data and farm data collection standards (Box et al., 2015; Amiri-Zarandi et al., 2022); suffice to say that this is a work in progress. A more detailed discussion around policy and other enabling factors can be found in Hansen et al. (2022).

2 The evolution from precision agriculture to smart farming

From an extensive farming perspective, the first systemic foray into digitalised agriculture was through precision agriculture (PA) which 'kicked off' in the 1990s with the 'switching-off' of selective availability in the U.S. global positioning systems (GPS) satellite positioning capability. Making the resulting metre-resolution positioning accuracy available in the public domain spawned a flurry activity in Australia, in sync with the rest of the world, with initial foci in the grains sector around harvester-mounted yield monitoring ('yield monitors'), on-ground soil surveying (e.g. the now ubiquitous EM-38) and airborne and satellite remote sensing (Cook and Bramley, 1998; Bramley and Trengove, 2013).

The broader concept of 'smart farming' was introduced in Australia only relatively recently (~20 years ago) on the back of the emergence of Agriculture 4.0 and the Internet of Things (IoT), and with the launch of the National Broadband (telecommunications) Network (NBN) (Lamb, 2013; Taylor et al., 2013; Mark et al., 2016; Holmes et al., 2020). Smart farming seeks to bring together constituent sensing, monitoring and recording elements into a whole-farm management and operations perspective. In other words, we are talking about digital agriculture enabled and enacted in the context of an entire farm. But are we achieving smart farming in the true sense of the word 'smart'? Are smart farms just a collection of data sources with a focus on the technology? This concept of aggregated technologies, namely 'SMART farms' was originally introduced by Lamb (2013) where SMART is merely an acronym for Sustainable, Manageable and Accessible Rural Technologies.

A whole farm can be considered an ecosystem of 'things' whereby the IoT is really the 'internet of assets' - all of which are generating pieces of information (in digital form) requiring aggregation, integration and ultimately decision-making (Figs 1 and 2). The concept of a smart farm as introduced in 2013 (Taylor et al.,

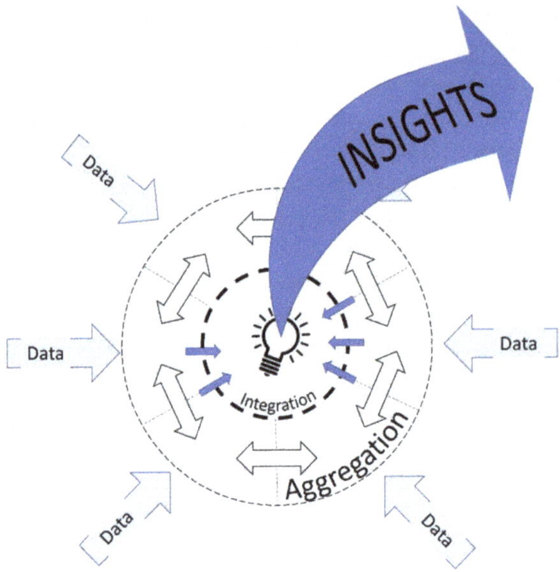

Figure 1 A conceptual relationship between data-generating devices (IoTs) on farms and the practice of smart farming. The heavy-dashed ring delineates the boundary between the data side and the decision/actions side- and this presently is co-located at the present boundary of data aggregation and data integration.

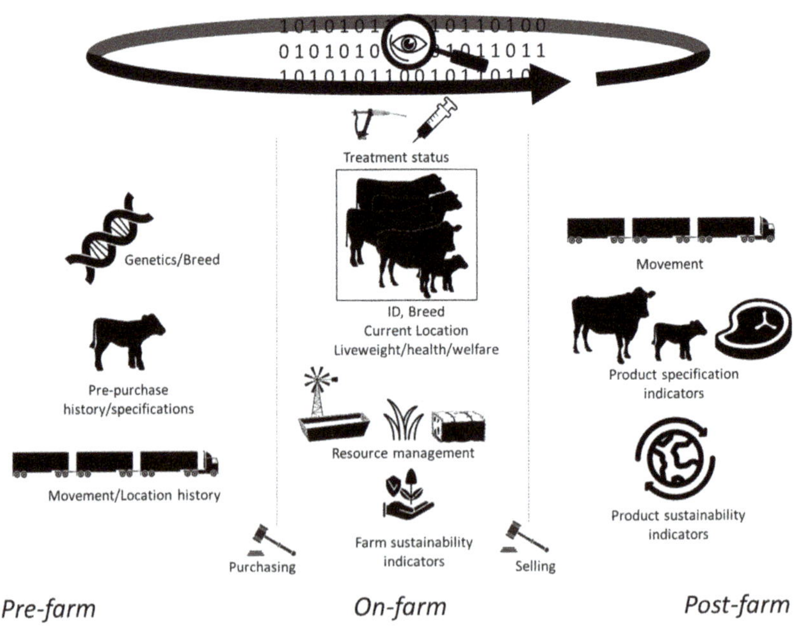

Figure 2 Typical data components of an extensive livestock production system in Australia.

2013) elevates the simple, one-dimensional situation, such as when an optical sensor responding to the signature of a green plant in an otherwise bare fallow field directly actuates the nozzle of a boom sprayer. This is where cyber-physical systems such as those outlined by Zhou et al. (2020) exist. However, unlike this all-digital example, smart farming still involves human-centric decision-making on the basis of a whole array of input signals.

Figure 1 depicts the relationship between data and smart farming. Data from a multitude of sources originates from individual sensors and it may stream in either on its own or may be aggregated. The ideal scenario is that rather than just aggregating data into a viewable dashboard, the data be rendered fully into a digital twin and then digitally integrated into the decision-making, or even automation framework (algorithms) (e.g. Lemphane et al., 2022).

Ideally, this could involve the integration of many and varied sources of digital information and allied insights 'at the digital side', prior to human consumption, with perhaps a digital dashboard and one or more decision support systems actively engaged. However, in practice, despite the promises of artificial intelligence (AI) and machine learning (ML) to 'do the work for us', and the notion that digital twins will somehow make it all easier for us to automate farming (Lemphane et al., 2022), and noting the substantial increase in R&D to support such interest in the context of livestock production systems (e.g. Bao and Xie, 2022; Lemphane et al., 2022), what we are seeing today is that integration is still occurring at the human side. 'Cognitive AI' is in its infancy (Zhao et al., 2022) and there is much work ahead to capture and codify 'human' on-farm decision making in AI systems. Multiple sources of digital information are being viewed by people and decisions are being undertaken inside people's heads. If there were to be a boundary delineating the 'digital side' (enabling data) and the 'human' or 'post-decision' side (enacting smart farming), it lies roughly as depicted by the heavy dashed ring in Fig. 1.

3 The adoption of digital agriculture and smart farming

Despite an increase in breadth of precision agriculture over time, and the attendant elements of digital agriculture that either support PA or integrates the outputs of PA into smart farming (Shepherd et al., 2020; Neethirajan and Kemp, 2021), the pace of adoption of smart farming in Australian farming systems remains slow. In assessing impediments to adoption of digital agriculture, much work to date has focused on the value proposition as considered by individual data/decision elements, individual producers or individual value chain actors. Cook et al. (2021) has provided an excellent review of how the lessons learned from PA are applicable to the broader realm of digital agriculture. At this level, adoption remains constrained by perceptions of high establishment and recurrent costs, lack of applicability and user accessibility and perceptions of technological risk (McFadden et al., 2022). Furthermore, in terms of integration of data and its role in farm decision-making, farmers still harbour reservations about overly relying on outputs of software. Many tend to question whether they can trust the outputs (Rotz et al., 2019). So as of today, smart farming remains SMART Farming with multiple data sources, including IoT coming to the farm manager for integration and conversion into insights and, ultimately, action.

To complete the scene setting for the Australian context, and this is a particular constraint of note, the large geographic size and the relative sparsity of populations in rural and regional areas creates challenges for telecommunications, in particular mobile telecommunications (Lamb, 2017a). Indeed, the Australian Broadband Advisory Council (2021) report, as subsequently published by Waters and Koch (2022), identified a 'connectivity threshold' where connectivity becomes an enabler of digital agriculture, rather than a drag on it? Here reliable connectivity was defined as '... connectivity across [...] paddocks [is] sufficient to use connected machinery and sensors, which can vary from narrowband to broadband depending on how 'bandwidth intensive' the apps are'. Their conclusion: 'So far, [in Australia] the threshold has not been consistently or reliably crossed, and this is holding back the adoption of digital agriculture'. It is only recently that policy settings, public and private initiatives, and a growing number of '2nd tier telecommunications providers' (Lamb, 2017a) are moving in recognition of the critical influence that reliable, ubiquitous telecommunications (in its many forms) has on digital innovations on farms (Lamb, 2017b, 2018; Hay and Eagle, 2021; Adams et al., 2022). While the telecommunications challenges faced by many Australian producers has influenced, and in many cases impeded, the adoption of digital innovations on farm and digital inclusion more broadly (Marshall et al., 2020), it has also influenced the type of technologies appearing on our smart farms, as revealed in the following sections.

4 The role of Internet of Things in extensive livestock production systems

The value proposition of digital agriculture, and the accompanying digitalisation of farming practice into smart farming will be empowered through the emergence of 'connected' farming. Typically taken to mean connected via the internet, here 'connectivity' means something else: the linking of food production through sustainability and value chains. For example, technology could support moves toward sustainability by connecting on-farm activities to changes in natural capital with producers rewarded through data-enabled land valuation or carbon financing (Cook et al., 2021). Alternatively, benefit accrues through traceability, coordination and value distribution for producers having 'high levels' of commercial integration along supply chains (Giua et al., 2022). Of course, connectivity adds complexity to on-farm production systems and this demands, in addition to tactical digital decision-making tools, integrative tools to assist in longer-term planning factoring in those complexities (Cook et al., 2021).

As far as Australia's $22 billion cattle and sheep sector is concerned (Australian Bureau of Statistics, 2023), adoption of digital technologies is complex and largely driven by traceability, product quality and differentiation (accreditation), good animal welfare practices (including accreditation) (Manning et al., 2021; Jackson and Cook, 2022; Schulz et al., 2022), drought resilience including feed base management (Henry et al., 2012; Bell et al., 2014; Lockie et al., 2020; Bowen and Chudleigh, 2021) and carbon neutrality (Mayberry et al., 2019).

Figure 2 depicts some of the important digital elements of an Australian extensive livestock enterprise. It should be noted that the 'T' in IoT, namely 'things', includes individual animals, again following the earlier notions that IoT is really the IoA ('A' being 'assets'). While the digitalisation of many of these elements is already advanced (e.g. Neethirajan and Kemp, 2021; Aquilani et al., 2022), including in Australia (Meehan, 2019), they are typically managed individually as data 'silos'. On the farm, however, many of the elements as related to herd inventory and status, as well as resource management, and now sustainability indicators are becoming integrated thanks to IoT devices and related services. But there is much work yet to be done. These will be the focus of the ensuring sections.

5 Technologies for livestock identification and tracking

The ultimate management unit in a livestock production system is the individual animal. Until recently, individual animal identification and subsequent traceability in extensive livestock production has been facilitated through the use of passive, RFID ear tags and the somewhat disconnected links in the Australian red meat integrity system (e.g. as described by Desmarchelier et al., 2007; Girish and Barbuddhe, 2020; McKinna and Wall, 2020). While principally designed with traceability and biosecurity in mind (e.g. as illustrated in Wilson et al., 2019; Girish and Barbuddhe, 2020; Seitzinger et al., 2022), the National Livestock Identification Scheme (NLIS) ear tags are also the primary identifier used on-farm by many data management platforms associated with management interventions such as routine or responsive weighing, drafting, drenching/dosing to name a few.

Tag retention is high as livestock progress through their lifecycle, although according to Meat and Livestock Australia, 'The 2018–9 traceability of 96.04% falls short of the 97% target' (Meat and Livestock Australia, 2021). There is consequently an ongoing interest in alternative means of animal identification in Australia including implantable devices (Meat and Livestock Australia, 2021), DNA-based technology (Banks et al., 2022) and facial recognition (Hossain et al., 2022). A useful review of many of these methods in the context of Australian extensive livestock production is given by Cosby et al. (2021).

Over the last decade, significant developments in animal mounted sensors, including those that incorporate location/position via global navigation satellite systems (GNSS) is offering the opportunity to augment passive RFID tags, which support events-based recording of animal interventions, with continuous measurement and monitoring, and even control of animals (Trotter, 2019; Windsor, 2021; Trotter et al., 2022; Lyons and Lomax, 2022).

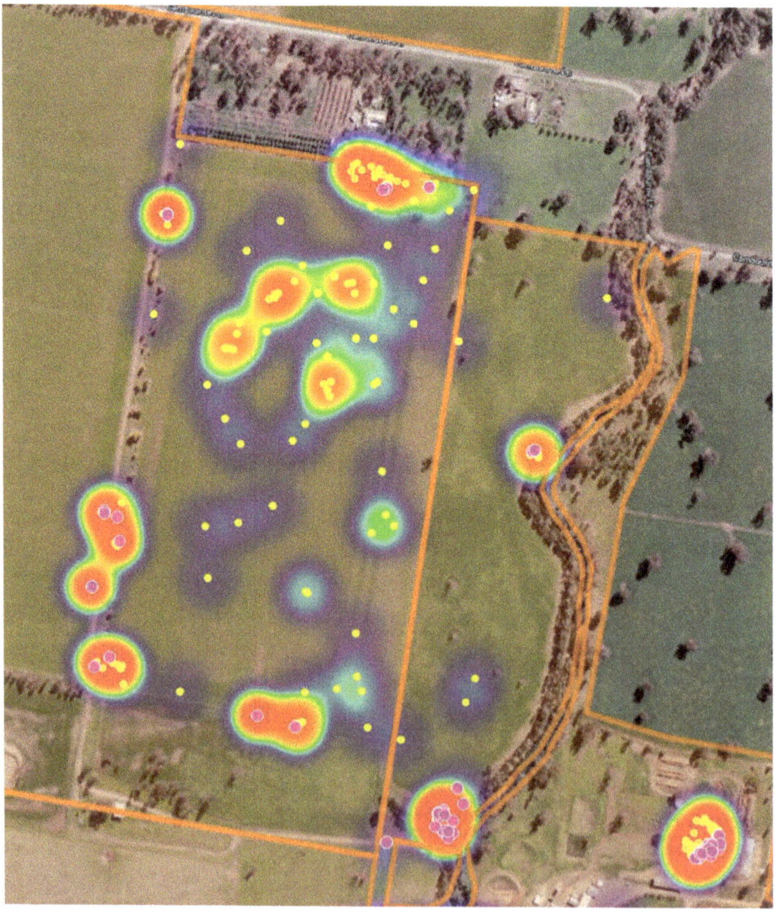

Figure 3 A livestock location and activity 'heat-map' as generated by Ceres Tag® (https://cerestag.com) from the Charles Sturt University Global Digital Farm (www.csu.edu.au/global-digital-farm).

Locating individual animals in an extensive farming system is not only the simplest outcomes but, in terms of farm operations, perhaps the most impactful outcome of on-animal sensors. Faced with the challenges of battery life and, at the time, a lack of maturity in cloud-based storage and analytical systems, initial demonstrations of livestock geolocation relied, somewhat paradoxically on collar mounted, store-on-board GNSS systems (e.g. Trotter et al., 2010a). Fast forward to today, and there are a range of energy efficient (including solar powered) ear tag livestock tracking systems for Australian extensive livestock producers available on the market, such as Ceres Tag® (https://cerestag.com; Campbell et al., 2022 and Fig. 3), Smart Paddock® (https://www.smartpaddock.com/) and mOOvement® (https://www.moovement.com.au/), to name a few.

6 Internet of Things applications in animal health and welfare

From an R&D perspective in this space, what stands out in Australia is the fact that much of the formative research in the use of on-animal sensors and devices has been carried out very much with extensive farming applications and challenges in mind. Examples include identification, traceability and market access (Fogarty et al., 2019; Cosby et al., 2021), general activity and welfare states (Barwick et al., 2018a,b, 2020; Fogarty et al., 2019, 2020a), feeding behaviour, preferencing and intake and water intake (Tomkins and O'Reagain, 2007; Thomas et al., 2008; Roberts et al., 2015; Greenwood et al., 2017; Manning et al., 2017; Williams et al., 2020), shelter use (Lamb et al., 2010; Taylor et al., 2011), rumination (Chang et al., 2022c), parturition (Dobos et at., 2014; Fogerty et al., 2020b,c, 2021; Chang et al., 2020, 2022a,b), post-partum parent-offspring pairing and interactions (Swain and Bishop-Hurley, 2007), illnesses like '3-day sickness' (Bovine ephemeral fever) (Tobin et al., 2020), parasite burden (Falzon et al., 2013), wild dog predation (Manning et al., 2014; Evans et al., 2022) and livestock theft (Trotter et al., 2018; Mulrooney et al., 2022).

In terms of device form factor, the majority of the sensors reported above are already in ear tag form, with some utilising halter or collar-mounted sensors (albeit many of them ear tags). In terms of devices available for commercial-scale deployment in extensive farming systems, ear tags dominate with current market offerings unchanged from those listed in the comprehensive review of Aquilani et al. (2022) (Table 2).

Again, with extensive farming in mind, Australia is considered a world leader in virtual fencing (Campbell et al., 2021). Here the interaction with the virtual fence boundary is facilitated via a collar worn by individual animals. Applications range from effecting control (Bishop-Hurley et al., 2007; Campbell et al., 2019, 2021, 2022) to excluding animals from sensitive areas (Campbell et al., 2018, 2020, 2022)

Substantial work has also been undertaken in understanding the nexus of what technology can do (i.e. pragmatic physics), how animal responds relative to cues and their welfare (Lee et al., 2009, 2018; Kearton et al., 2019; Lee and Campbell, 2021). Moreover with Australian conditions and context in mind with this raft of IoT opportunities comes the necessity to consider, where necessary harmonise and even propose, legal and compliance frameworks of emerging industry practice, animal health and wellbeing and community expectations (Lee et al., 2018; Manning et al., 2021).

7 Combining Internet of Things devices: live weight and yield mapping

Regular, physical measurements of cattle live weight remains the primary point of truth when it comes to gauging not only the day-to-day effectiveness of current feed regimes (e.g. stocking rate, feed composition, etc.) but also in terms of planning/scheduling turn off and replacement/replenishment as well as for deciding when/how to send them onto the market (sale yards, feedlots and/or processing). Of course, manual handling and weighing of cattle is time consuming, can adversely affect animals and is disruptive to other farm operations. While the subject of research and development for more than 40

Figure 4 An Optiweigh® front-feet weighing system deployed on the Charles Sturt University Global Digital Farm (www.csu.edu.au/global-digital-farm).

years (for example Smith and Turner, 1974; Filby et al., 1979; Long et al., 1991), the use of walk-over weighing systems (WoW) either in the animal handling facilities, or better still for extensive farming, out in paddocks is now relatively common (Leigo et al., 2012; Trotter, 2013; Menzies et al., 2018; Aquilani et al., 2021). More recently, in-paddock, front-foot weighing systems like the Optiweigh® system (https://www.optiweigh.com.au/) (Fig. 4), are finding increasing use in commercial extensive farming systems.

The integration of regularly assessed live weight with animal tracking data has been investigated with the view of generating paddock-based yield maps. The tactic here is that providing producers with such maps may capture the interest/imagination for extensive livestock producers to engage in precision pasture management; not dissimilar to the same way that grains yield maps (generated from GNSS-yield monitor equipped harvesters) spawned

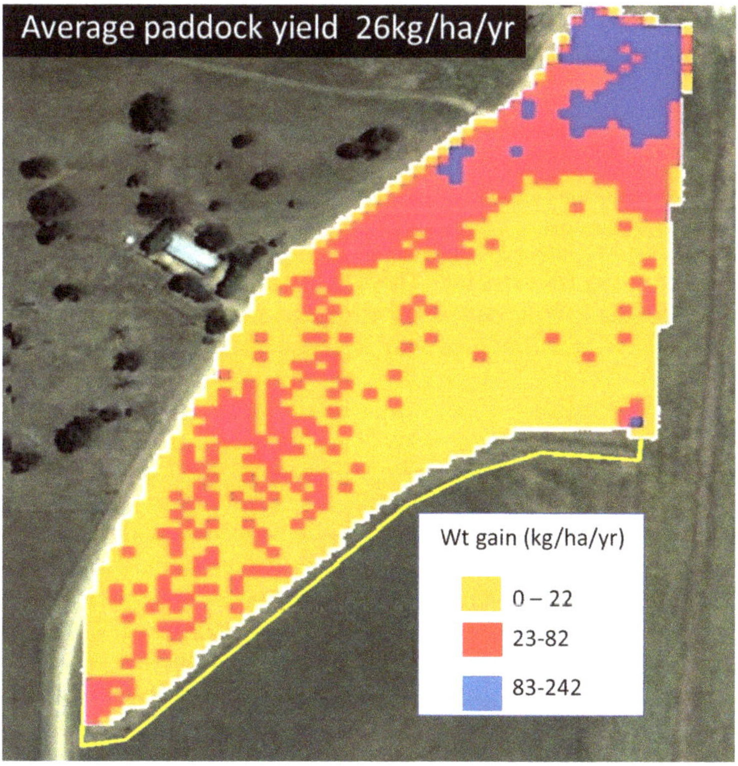

Figure 5 A sheep-meat yield map for 20 mixed aged Merino ewes over a period of 43 days using a combination of live weighing and GNSS-enabled animal movement data. Weights (Wt) are extrapolated to a growing cycle of 1 year. Original unpublished data and subsequent analysis courtesy of Zac Economou, Derek Schneider and the author, University of New England SMART Farms (www.une.edu.au/smartfarms), 2015.

variable-rate (precision) fertiliser management during the late 1990s–early 2000s (e.g. Blackmer and White, 1998; Adams et al., 2000; Lamb et al., 2008; Hedley, 2015).

An example of a livestock 'yield map' that can be generated from a combination of WoW and GNSS tracking data is illustrated in Fig. 5 (generated from original 2015 unpublished data courtesy of Zac Economou, Derek Schneider and the author). This map has been generated for 20 mixed aged Merino ewes over a period of 43 days using a combination of live weight gain and GNSS-enabled animal movement data. Conducted on the University of New England SMART Farms (www.une.edu.au/smartfarms), and assuming a dry-stock-equivalent (DSE) rating of 1 DSE per dry ewe, sheep were run at an average stocking rate of 5.4 DSE per ha. Raw tracking data was first converted into a livestock residence index (LRI) following Trotter et al. (2010b), then segmented into grazing and non-grazing epochs following the method of Putfarken et al. (2008), the former ascribed to weight gain assuming minimal weigh loss during non-grazing activity. Weight gain was apportioned into wool growth, fixed rate calculated to be 5.75 g/head/day (Wheeler et al., 1977) and an increase in live weight (muscle and fat). Gut-fill was ignored as an influence on live weight as the focus was on spatial changes in weight gain, which would have averaged out across the paddock. Live-weight gains were then finally extrapolated to a growing period of 1 year.

8 Combining Internet of Things devices: water intake and water infrastructure failure alerting

On their own, water pressure, flow and volume sensors placed on water points and on feeder pipes are valuable for monitoring gross water consumption and water system performance (e.g. leaks, burst pipes, etc.). Figure 6 depicts a portion of the farm-wide, Farmbot® water monitoring system on the Charles Sturt University Global Digital farm (www.csu.edu.au/global-digital-farm), and the recorded data as rendered on the FarmBot® data monitoring system. Data includes pressure, flow rates and volume of water for individual livestock water troughs.

There is also a significant link between animal water consumption and health and welfare (e.g. von Keyserlingk et al., 2016; Wagner and Engle, 2021). It is not surprising that the power of data analytics is now being turned towards inferring animal health and welfare from the data such as Fig. 6. A combination of on-animal sensors and water point sensors offers the means of monitoring both water intake, as an indicator of animal health and productivity, animal health as well as for alerting to failures in the water points themselves (e.g. Williams et al., 2020; Tang et al., 2021; Tobin et al., 2021).

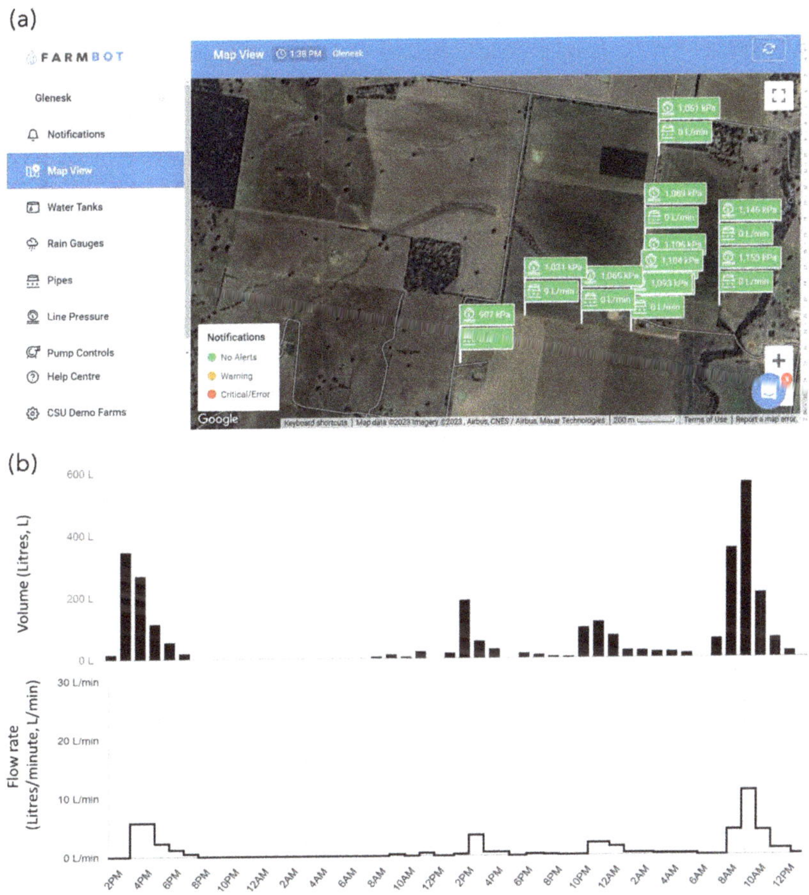

Figure 6 (a) Water feeder pipe pressure and water pressure, flow rates and volumes at individual livestock water troughs on the Charles Sturt University Global Digital Farm (www.csu.edu.au/global-digital-farm). (b) Diurnal flow of water consumed by 70 x ~300 kg Angus steers on one of the fields over a 48-h period. Data provided by Farmbot® (www.farmbot.com.au) and Nelson West (CSU Global Digital Farm).

9 Combining Internet of Things devices: feed base and stocking rate management

Given the critical links to animal health and welfare and productivity, ultimately one of the highest-priority strategic and tactical decisions made by livestock producers concerns management of the feed-base (e.g. Dicker et al., 1998; Bell et al., 2008; Bentley et al., 2008; Moore et al., 2009; Moore et al., 2021; Bell and Sangster, 2022). This is becoming even more so with the challenges of climate change (Ghahramani et al., 2019; Cullen et al., 2021; Williams et al., 2021).

Predominant work to date for pasture evaluation is via remote sensing methods including satellite (optical and radar) and, more recently, drones (Crabbe et al., 2019a, 2019b; Gargiulo et al., 2020; Alvarez-Hess et al., 2021; Chen et al., 2021; Pringle et al., 2021; and references therein). Recently, Meat Livestock Australia (MLA) has released a remote sensing-based feed-base monitor available to all Australian livestock producers free of charge (Meat and Livestock Australia, 2023a).

Very little has been reported around using IoT-based devices per se. However, of relevance here to potential future IoT manifestations and deployments is the growth in interest and activity in Australia in using hand-held and mobile devices for quantifying the feed-base. A review of the many forms of portable (human or vehicle) mechanical devices is given in Trotter et al. (2010c). What is of particular interest and relevance to the context of this chapter are proximal active and passive optical sensors as these are amenable to use as fixed, sentinel IoT devices quantifying and predicting characteristics of feed-base. The use of portable active and passive optical sensing devices for estimating pasture biomass in a variety of pasture systems and of varying compositions is discussed and illustrated in Trotter et al. (2010d), Rahman et al. (2014), Schaefer et al. (2014), Mundava et al. (2015), Andersson et al. (2017), Rahman and Lamb (2017) and Alam et al. (2018) to name a few.

With direct reference to IoT-type deployments, however there is far less to report. Over in New Zealand, and with dairy feed-base management in mind, Milsom et al. (2019) describe the use of a static, pole-mounted time-of-flight optical depth sensor to quantify the height (and growth rate through temporal changes in height) of perennial ryegrass (*Lolium perenne*). While encouraging, noting 'The devices were at least as accurate at estimating DM as a RPM' (rising plate meter), the authors did allude to plans to upgrading their system with moisture and temperature probes. Meanwhile, Rahman et al. (2014) investigated the use of multiple sensors to quantify pasture growth using a light-use efficiency model. A salient feature of this work is the use of *in situ* active optical sensing of the fraction of absorbed photosynthetic radiation (fAPAR) by the pasture canopy, a time-domain reflectometer to estimate the soil moisture content (expressed as a moisture index - MI) as well as a micro-climate weather station to derive local temperature (expressed a temperature index - TI). From this combination of sensors, it was possible to estimate the growth rate of the candidate pasture species - in this case tall fescue (*Festuca arundinacea*) with a root mean square error of 1.68 kg/ha day. Indeed, in the 2014 paper, the authors conclude that 'When coupled with the development of low-cost sensor networks and mobile device applications (mobile 'apps')–the so-called 'Internet of Things' on farms–it is not difficult to imagine future, instrumented, farming landscapes providing real-time management data concerning pasture (and crop) growth rates as an aid to management decisions.' Figure 7 illustrates

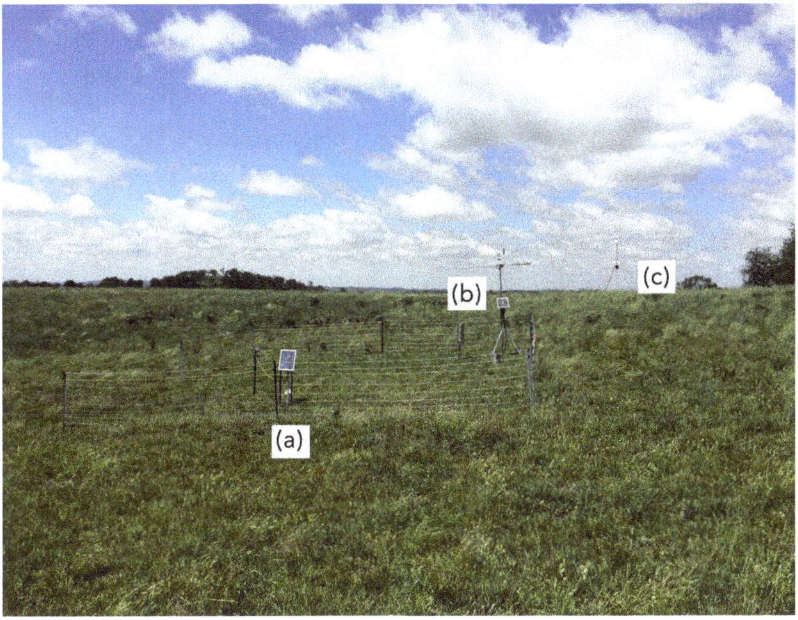

Figure 7 A combination of (a) active optical pasture sensor, (b) integrated weather station and soil moisture probe and (c) store-and-forward telemetry node as part of the UNE SMART Farms (www.une.edu.au/smartfarms) sensor network for monitoring pasture growth rates.

an integrated pasture growth sensor array located on the University of New England SMART Farms.

Since this early work in 2014–5, MLA, in partnership with Australian companies Precision Pastures (https://www.precisionpastures.com.au/) and ICT International (https://au.ictinternational.com/), has recently embarked on producer trials of similar technology aimed at improving sustainability (Meat and Livestock Australia, 2023b) (Fig. 8).

10 The rise of demonstration smart farms for extensive livestock production

There is little doubt that agtech, and in particular connected things will play an increasingly important, enabling role in extensive livestock production systems in improving productivity, safety, workflow, sustainability and resilience in the face of climate change, and market access. But the key 'enablers of such enablement' remain the people (Leonard et al., 2017). Demonstration farms, and in particular demonstration smart farms, featuring, amongst other things, IoT devices in action are a key platform for education, outreach and inspiration for our producers.

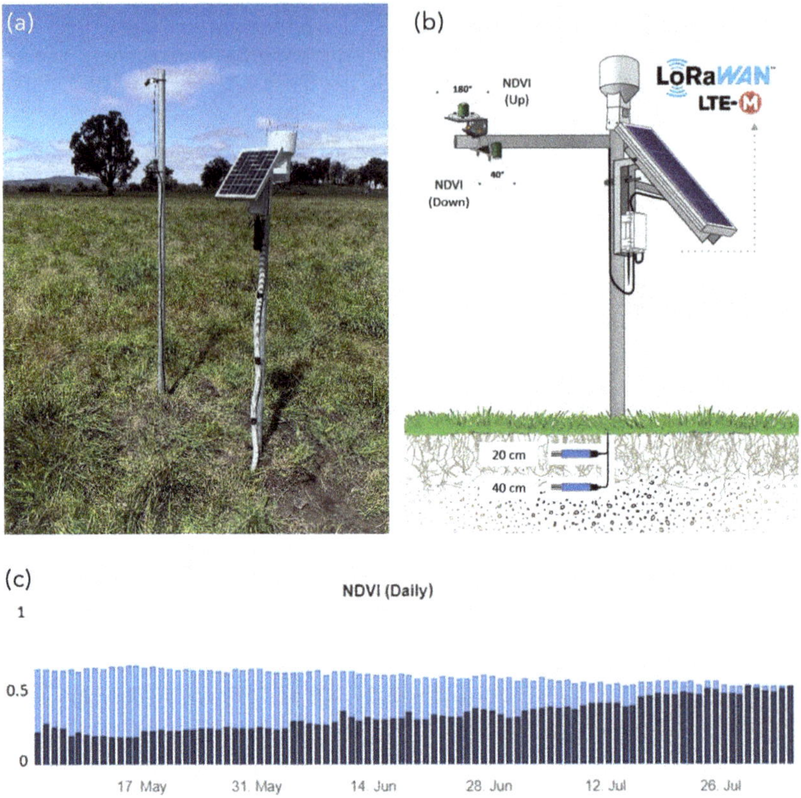

Figure 8 (a) Photograph and (b) schematic of an integrated pasture normalised difference vegetation index (NDVI) sensor, weather station and soil moisture array for observing pasture biomass evolution over time. Automatically sampled NDVI data is depicted in (c) for two different pasture types (light blue and dark blue columns). Imagery and data courtesy of Dr Peter Cull, ICT International (https://au.ictinternational.com/).

It is reasonable to end this chapter where it all began - or, at least where Australian smart farming began. The concept of SMART ('Sustainable, Manageable and Accessible Rural Technologies') Farms was first brought to life by the University of New England (UNE) (Armidale, NSW) in the mid-2000s' as part of the National Broadband Network rollout (NBNCo, 2006). Through a raft of technology demonstrators (livestock tracking, remote infrastructure monitoring and video connectivity), the UNE SMART Farm was initially focussed on demonstrating the potential value of broadband connectivity to the business and lifestyle of farming. Awarded a Regional Development Australia 'Innovation of the Year Award' in 2014 (RDA, 2014), the UNE SMART Farm also demonstrated an even more important role, albeit not fully recognised at the time, and a niche one at that. University-hosted smart farms can bring together

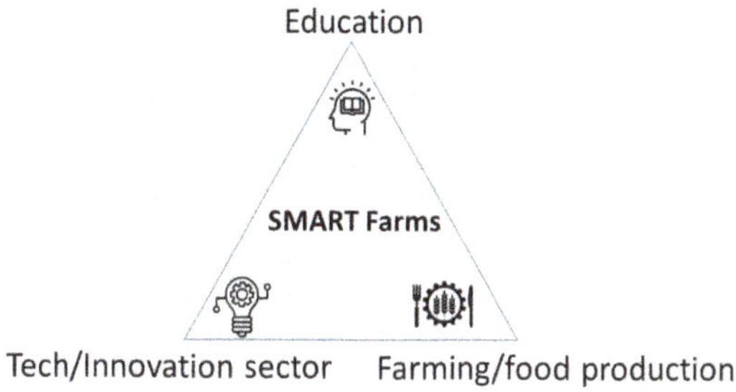

Figure 9 Smart Farms hosted in regional universities have the potential to foster telecommunications-enabled innovation.

Figure 10 An IoT-fuelled, data-rich environment. Livestock on the Charles Sturt University Global Digital Farm (www.csu.edu.au/global-digital-farm), each taking water from FarmBot® sensorised troughs, with a nearby Optiweigh® front-foot weigher and each is wearing a Ceres Tag® GNSS-tracking ear tag. Pasture biomass is being monitored remotely using a CiboLabs PastureKey ®.Photos courtesy of Jonathon Medway (reproduced with permission) and the author.

Education/Outreach, Telecommunications-enabled technology innovation and Farming/Food Production stakeholders (Fig. 9) together in a single, lasting framework.

At the present time, two of Australia's premier regional Universities: University of New England and Charles Sturt University located in regional NSW have fully-functioning smart farms. Both are acknowledged national leaders in the provision of undergraduate and postgraduate training in agriculture, rural and environmental sciences, as well as in STEM-related disciplines (an important 'co-ingredient' in supporting technology and data innovation). Both universities have substantial rural property holdings adjacent to campuses (UNE, Armidale:

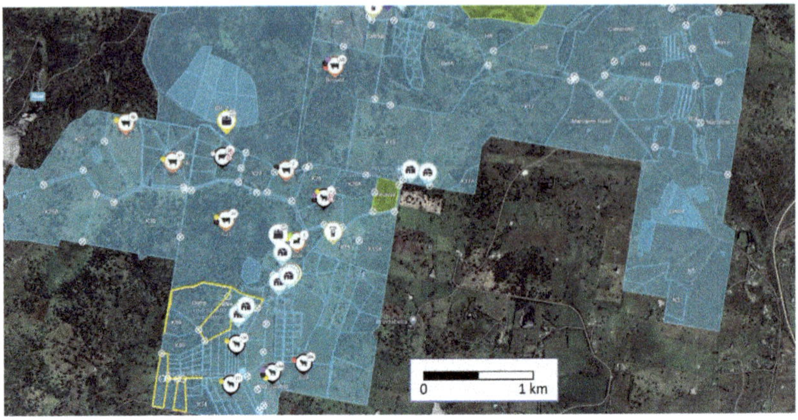

Figure 11 Presently, extensive smart farming in Australia is about collecting and collating data from a multitude of sources and rendering them in a single location/platform. This example is a portion of the University of New England (Australia) SMART Farm (www.une.edu.au/smartfarms) dashboard using the AgriWebb® platform (www.agriwebb.com). Covering ~2000 ha of prime livestock production land, records include paddock carrying capacity and actual stocking rate (E-ewes, W-whethers), water (flow) pipe sensors, water tank level sensors and weather stations (photo by author).

2900 ha 'UNE SMART Farms'; CSU, Wagga Wagga, 1800 ha 'Global Digital Farm') and both have substantive and substantial onsite relationships with state government Departments of Primary Industries (or their equivalent) as well as local and regional producer groups. Charles Sturt University Global Digital Farm has a focus on mixed farming comprising broadacre and extensive livestock production (Fig. 10) whereas University of New England has a focus on both intensive and extensive livestock production as part of an integrated whole of lifecycle livestock production system (Fig. 11).

11 Conclusion

University-hosted smart farms are an opportunity to showcase the latest technologies for improving productivity, sustainability and support for Australian farms. In addition to offering a safe environment to innovate and a realistic testing ground to validate operating conditions and formative assumptions, they provide the context for problem-based experiential learning for undergraduates and for research and community activities. University, college and agency-hosted smart farms are valuable resources that can help solve real-world problems and enhance the future of agriculture. It is not surprising that an initiative to form a Global Smart Farm Network has recently been announced (https://www.longy.com.au/2023/11/15/global-smart-farm-network).

12 Acknowledgements

The author wishes to acknowledge the support of Food Agility Co-operative Research Centre (CRC) Ltd, funded under the Commonwealth Government CRC Program. The CRC Program supports industry-led collaborations between industry, researchers and the community.

The author wishes to acknowledge the contributions of CSU Global Digital Farm Director Jonathon Medway and Senior Technical Officer Nelson West, as well as UNE SMART Farms Senior Technical Officer Derek Schneider for numerous helpful discussions and assistance while assembling material relevant to this chapter.

Specific mention of companies, products or services by the author does not constitute an endorsement.

13 Where to look for further information

Details of the breadth of research and development activities conducted by Food Agility CRC can be found at www.foodagility.com.

The Global Digital Farm (GDF) initiative (www.csu.edu.au/global-digital-farm) located at Wagga Wagga, NSW Australia, is a partnership between Charles Sturt University and Food Agility CRC (www.csu.edu.au/global-digital-farm).

The University of New England (Australia) SMART Farms initiative (www.une.edu.au/smartfarms) is located at Armidale, NSW Australia and is acknowledged as one of world's first commercial scale 'smart farms'. It was established in 2007.

14 References

ABARES (2021). Snapshot of Australian agriculture 2021. Australian Bureau of Agricultural and Resource Economics and Sciences, Canberra. https://doi.org/10.25814/rxjx-3g23.

ABARES (2023). Snapshot of Australian agriculture 2023. ABARES Insights, Canberra, March. https://doi.org/10.25814/rk1z-qm36.

Adams, M. L., Cook, S. and Bowden, J. W. (2000). Using yield maps and intensive soil sampling to improve nitrogen fertiliser recommendations from a deterministic model in the Western Australian wheatbelt. *Australian Journal of Experimental Agriculture* 40(7), 959–968.

Adams, P., Inglis, S. and Proctor, J. (2022). The broadband futures forum: Regional connectivity and shared infrastructure in NSW and New Zealand. *Journal of Telecommunications and the Digital Economy* 10(3), 1–13.

Alam, M. S., Lamb, D. W. and Rahman, M. M. (2018). A refined method for rapidly determining the relationship between canopy NDVI and the pasture evapotranspiration coefficient. *Computers and Electronics in Agriculture* 147, 12–17. https://doi.org/10.1016/j.compag.2018.02.008.

Alvarez-Hess, P. S., Thomson, A. L., Karunaratne, S. B., Douglas, M. L., Wright, M. M., Heard, J. W., Jacobs, J. L., Morse-McNabb, E. M., Wales, W. J. and Auldist, M. J. (2021). Using multispectral data from an unmanned aerial system to estimate pasture depletion during grazing. *Animal Feed Science and Technology* 275, 114880.

Amiri-Zarandi, M., Hazrati Fard, M., Yousefinaghani, S., Kaviani, M. and Dara, R. A. (2022). A platform approach to Smart farm information processing. *Agriculture* 12(6), 838. https://doi.org/10.3390/agriculture12060838.

Andersson, K., Trotter, M., Robson, A., Schneider, D., Frizell, L., Saint, A., Lamb, D. and Blore, C. (2017). Estimating pasture biomass with active optical sensors. *Advances in Animal Biosciences* 8(2), 754–757. https://doi.org/10.1017/S2040470017000838.

Annosi, C., Brunetta, F., Capo, F. and Heideveld, L. (2020). Digitalization in the agri-food industry: The relationship between technology and sustainable development. *Management Decision* 58(8), 1737–1757. https://doi.org/10.1108/MD-09-2019-1328.

Aquilani, C., Confessore, A., Bozzi, R., Sirtori, F. and Pugliese, C. (2022). Review: Precision livestock farming technologies in pasture-based livestock systems. *Animal* 16(1), 100429. https://doi.org/10.1016/j.animal.2021.100429.

Australian Broadband Advisory Council (2021). Report. Available at: https://www.infrastructure.gov.au/sites/default/files/documents/agri-tech-expert-working-group.pdf. (Accessed April 2023).

Australian Bureau of Statistics (2023). Value of agricultural commodities produced 2021–2022, Australia. Available at: https://www.abs.gov.au/statistics/industry/agriculture/value-agricultural-commodities-produced-australia/latest-release. (Accessed April 2023).

Banks, R., Miller, S., van der Werk, J., Ball, A. and Woolsaton, R. (2022). Genotyping for individual animal ID. Opportunities for Utilizing DNA-Based Technologies to Enhance Traceability and Profitability in the Australian Red Meat Industry. Final Report. Australia: Food Agility Cooperative Research Centre. 100 pages. ISBN: 978-0-6454652-4-2. Available at: https://assets-global.website-files.com/5f4f19737a6ae318c84e362c/646eaecabb25b65def87b041_FA073_Traceability%20Report_Final.pdf. (Accessed May 2023).

Bao, J. and Xie, Q. (2022). Artificial intelligence in animal farming: A systematic literature review. *Journal of Cleaner Production* 331, 129956. https://doi.org/10.1016/j.jclepro.2021.129956.

Barwick, J., Lamb, D. W., Dobos, R., Welch, M. and Trotter, M. (2018a). Categorising sheep activity using a tri-axial accelerometer. *Computers and Electronics in Agriculture* 145, 289–297. https://doi:10.1016/j.compag.2018.01.007.

Barwick, J., Lamb, D., Dobos, R., Schneider, D., Welch, M. and Trotter, M. (2018b). Predicting lameness in sheep activity using tri-axial acceleration signals. *Animals* 8(1), 1–16. https://doi.org/10.3390/ani8010012.

Barwick, J., Lamb, D. W., Dobos, R., Welch, M., Schneider, D. and Trotter, M. G. (2020). Identifying sheep activity from tri-axial acceleration signals using a moving window classification model. *Remote Sensing* 12(4), 1–13. https://doi.org/10.3390/rs12040646.

Beck, H. E., Zimmermann, N. E., McVicar, T. R., Vergopolan, N., Berg, A. and Wood, E. F. (2018). Present and future Köppen-Geiger climate classification maps at 1-km resolution. *Nature Scientific Data* 5, 180214. https://doi.org/10.1038/sdata.2018.214.

Bell, L. W., Robertson, M. J., Revell, D. K., Lilley, J. M. and Moore, A. D. (2008). Approaches for assessing some attributes of feed-base systems in mixed farming enterprises. *Australian Journal of Experimental Agriculture* 48(7), 789–798.

Bell, L. W., Hayes, R. C., Pembleton, K. G. and Waters, C. M. (2014). Opportunities and challenges in Australian grasslands: Pathways to achieve future sustainability and productivity imperatives. *Crop and Pasture Science* 65(6), 489–507.

Bell, A. and Sangster, N. (2022). Research, development and adoption for the north Australian beef cattle breeding industry: An analysis of needs and gaps. *Animal Production Science* 63(1), 1–40.

Bentley, D., Hegarty, R. S. and Alford, A. R. (2008). Managing livestock enterprises in Australia's extensive rangelands for greenhouse gas and environmental outcomes: A pastoral company perspective. *Australian Journal of Experimental Agriculture* 48(2), 60–64.

Bishop-Hurley, G. J., Swain, D. L., Anderson, D. M., Sikka, P., Crossman, C. and Corke, P. (2007). Virtual fencing applications: Implementing and testing an automated cattle control system. *Computers and Electronics in Agriculture* 56(1), 14–22.

Blackmer, A. M. and White, S. E. (1998). Using precision farming technologies to improve management of soil and fertiliser nitrogen. *Australian Journal of Agricultural Research* 49(3), 555–564.

Bowen, M. K. and Chudleigh, F. (2021). Achieving drought resilience in the grazing lands of northern Australia: Preparing, responding and recovering. *The Rangeland Journal* 43(3), 67–76.

Box, P., Simons, B., Cox, S. and Maguire, S. (2015). A data specification framework for the foundation spatial data framework. CSIRO, Australia.

Bramley, R. and Trengove, S. (2013). Precision agriculture in Australia: Present status and recent developments. *Engenharia Agrícola* 33(3), 575–588.

Campbell, D. L. M., Haynes, S. J., Lea, J. M., Farrer, W. J. and Lee, C. (2018). Temporary exclusion of cattle from a riparian zone using virtual fencing technology. *Animals* 9(1), 5.

Campbell, D. L. M., Lea, J. M., Keshavarzi, H. and Lee, C. (2019). Virtual fencing is comparable to electric tape fencing for cattle behavior and welfare. *Frontiers in Veterinary Science* 6, 445.

Campbell, D. L. M., Ouzman, J., Mowat, D., Lea, J. M., Lee, C. and Llewellyn, R. S. (2020). Virtual fencing technology excludes beef cattle from an environmentally sensitive area. *Animals* 10(6), 1069.

Campbell, D. L. M., Marini, D., Lea, J. M., Keshavarzi, H., Dyall, T. R. and Lee, C. (2021). The application of virtual fencing technology effectively herds cattle and sheep. *Animal Production Science* 61(13), 1393–1402.

Campbell, D. L. M., Bishop-Hurley, G. J., Lee, C. and Charmley, E. (2022). Developments in monitoring grazing behaviour and automated grazing management in extensive systems. In Berckmans, D. (Ed.), *Advances in Precision Livestock Farming*. Belgium: Burleigh Dodds Science Publishing Limited, 385–408. https://doi.org/10.19103/AS.2021.0090.14.

Chang, A. Z., Swain, D. L. and Trotter, M. G. (2020). Towards sensor-based calving detection in the rangelands: A systematic review of credible behavioral and physiological indicators. *Translational Animal Science* 4(3), 1–18. https://doi.org/10.1093/tas/txaa155.

Chang, A. Z., Fogarty, E. S., Swain, D. L., García-Guerra, A. and Trotter, M. G. (2022a). Accelerometer derived rumination monitoring detects changes in behaviour around parturition. *Applied Animal Behaviour Science* 247, 1–11. https://doi.org/10.1016/j.applanim.2022.105566.

Chang, A. Z., Swain, D. L. and Trotter, M. G. (2022b). A multi-sensor approach to calving detection. *Information Processing in Agriculture*, 1–20. https://doi.org/10.1016/j.inpa.2022.07.002.

Chang, A. Z., Fogarty, E. S., Moraes, L. E., García-Guerra, A., Swain, D. L. and Trotter, M. G. (2022c). Detection of rumination in cattle using an accelerometer ear-tag: A comparison of analytical methods and individual animal and generic models. *Computers and Electronics in Agriculture* 192, 1–12. https://doi.org/10.1016/j.compag.2021.106595.

Chen, Y., Guerschman, J., Shendryk, Y., Henry, D. and Harrison, M. T. (2021). Estimating pasture biomass using Sentinel-2 imagery and machine learning. *Remote Sensing* 13(4), 603.

Coelli, R. (2021). Natural resource management and drought resilience - Survey of farm practices. ABARES research report 21.12, Canberra. https://doi.org/10.25814/99n0-7q92.

Cook, S. E. and Bramley, R. G. V. (1998). Precision agriculture–opportunities, benefits and pitfalls of site-specific crop management in Australia. *Australian Journal of Experimental Agriculture* 38(7), 753–763.

Cook, S., Jackson, E. L., Fisher, M. J. (In Memoriam), Baker, D. and Diepeveen, D. (2021). Embedding digital agriculture into sustainable Australian food systems: Pathways and pitfalls to value creation. *International Journal of Agricultural Sustainability*. https://doi.org/10.1080/14735903.2021.1937881.

Cosby, A., Manning, J., Fogarty, E., Wilson, C., Islam, N. and Trotter, M. (2021). Assessing real time tracking technologies to integrate with identification methods and national traceability requirements. Final Report, V.RDA.2005. Meat and Livestock Australia. North Sydney, NSW 2059: Meat and Livestock Australia Limited. Available at: https://www.mla.com.au/globalassets/mla-corporate/research-and-development/final-reports/2021/v.rda.2005-final-report.pdf. (Accessed May 2023).

Crabbe, R. A., Lamb, D. W., Edwards, C., Andersson, K. and Schneider, D. (2019a). A preliminary investigation of the potential of sentinel-1 radar to estimate pasture biomass in a grazed pasture landscape. *Remote Sensing* 11(7), 872.

Crabbe, R. A., Lamb, D. W. and Edwards, C. (2019b). Discriminating between C3, C4, and mixed C3/C4 pasture grasses of a grazed landscape using multi-temporal Sentinel-1a data. *Remote Sensing* 11(3), 253.

Cullen, B., Harrison, M., Mayberry, D., Cobon, D., Davison, T. and Eckard, R. (2021). Climate change impacts and adaption strategies for pasture-based industries: Australian perspective. NZGA: Research and Practice Series 17, 139–148.

DAWE (2020). Delivering Ag2030. Australian Government Department of Agriculture, Water and the Environment, Canberra.

Desmarchelier, P., Fegan, N., Smale, N. and Small, A. (2007). Managing safety and quality through the red meat chain. *Meat Science* 77(1), 28–35. https://doi.org/10.1016/j.meatsci.2007.04.027.

Dicker, R. W., Ayres, J. F., Dobos, R. C., McPhee, M. J., Nandra, K. S. and Turner, A. D. (1998). The pasture feed-base for beef cattle production on the northern tablelands of New South Wales: A review. *Animal Production in Australia* 22, 181–184.

Dobos, R. C., Dickson, S., Bailey, D. W. and Trotter, M. G. (2014). The use of GNSS technology to identify lambing behaviour in pregnant grazing Merino ewes. *Animal Production Science* 54(10), 1722–1727. https://doi.org/10.1071/AN14297.

Evans, C. A., Trotter, M. G. and Manning, J. K. (2022). Sensor-based detection of predator influence on livestock: A case study exploring the impacts of wild dogs (Canis familiaris) on rangeland sheep. *Animals* 12(3), 1–11. https://doi.org/10.3390/ani12030219.

Falzon, G., Trotter, M. G., Schneider, D. and Lamb, D. W. (2013). Correlating movement patterns of merino sheep to faecal egg counts using global positioning system tracking collars and functional data analysis. *Small Ruminant Research* 111, 171–174. https://doi.org/10.1016/j.smallrumres.2012.09.001.

Filby, D. E., Turner, M. J. B. and Street, M. J. (1979). A walk-through weigher for dairy cows. *Journal of Agricultural Engineering Research* 24(1), 67–78.

Fogarty, E. S., Swain, D. L., Cronin, G. M. and Trotter, M. G. (2019). A systematic review of the potential uses of on-animal sensors to monitor the welfare of sheep evaluated using the Five Domains Model as a framework. *Animal Welfare Journal* 28(4), 407–420. https://doi.org/10.7120/09627286.28.4.407.

Fogarty, E. S., Swain, D. L., Cronin, G. M., Moraes, L. E. and Trotter, M. G. (2020a). Behaviour classification of extensively grazed sheep using machine learning. *Computers and Electronics in Agriculture* 169, 1–10. https://doi.org/10.1016/j.compag.2019.105175.

Fogarty, E. S., Swain, D. L., Cronin, G. M., Moraes, L. E. and Trotter, M. G. (2020b). Can accelerometer ear tags identify behavioural changes in sheep associated with parturition? *Animal Reproduction Science* 216, 1–13. https://doi.org/10.1016/j.anireprosci.2020.106345.

Fogarty, E. S., Swain, D. L., Cronin, G. M., Moraes, L. E., Bailey, D. W. and Trotter, M. G. (2020c). Potential for autonomous detection of lambing using global navigation satellite system technology. *Animal Production Science* 60(9), 1217–1226. https://doi.org/10.1071/AN18654.

Fogarty, E. S., Swain, D. L., Cronin, G. M., Moraes, L. E., Bailey, D. W. and Trotter, M. G. (2021). Developing a simulated online model that integrates GNSS, accelerometer and weather data to detect parturition events in grazing sheep: A machine learning approach. *Animals* 11(2), 1–22. https://doi.org/10.3390/ani11020303.

Fuglie, K. O. (2018). Is agricultural productivity slowing? *Global Food Security* 17, 73083. https://doi.org/10.1016/j.gfs.2018.05.001.

Gargiulo, J., Clark, C., Lyons, N., de Veyrac, G., Beale, P. and Garcia, S. (2020). Spatial and temporal pasture biomass estimation integrating electronic plate meter, planet cubesats and sentinel-2 satellite data. *Remote Sensing* 12(19), 3222.

Ghahramani, A., Howden, S. M., del Prado, A., Thomas, D. T., Moore, A. D., Ji, B. and Ates, S. (2019). Climate change impact, adaptation, and mitigation in temperate grazing systems: A review. *Sustainability* 11(24), 7224.

Girish, P. S. and Barbuddhe, S. B. (2020). Meat traceability and certification in meat supply chain. In *Meat Quality Analysis*. Editor(s): Ashim Kumar Biswas, Prabhat Kumar Mandal, Academic Press, 153–170, ISBN 9780128192337, https://doi.org/10.1016/B978-0-12-819233-7.00010-0.

Giua, C., Materia, V. C. and Camanzi, L. (2022). Smart farming technologies adoption: Which factors play a role in the digital transition? *Technology in Society* 68, 101869. https://doi.org/10.1016/j.techsoc.2022.101869.

Greenwood, P. L., Paull, D. R., McNally, J., Kalinowski, T., Ebert, D., Little, B., Smith, D. V., Rahman, A., Valencia, P., Ingham, A. B. and Bishop-Hurley, G. J. (2017). Use of sensor-determined behaviours to develop algorithms for pasture intake by individual grazing cattle. *Crop and Pasture Science* 68(12), 1091–1099. https://doi.org/10.1071/CP16383.

Hansen, B. D., Leonard, E., Mitchell, M. C., Easton, J., Shariati, N., Mortlock, M. Y., Schaefer, M. and Lamb, D. W. (2022). Current status of and future opportunities for digital agriculture in Australia. *Crop and Pasture Science* 74(6), 524–537. https://doi.org/10.1071/CP21594.

Hay, R. and Eagle, L. (2021). Marketing social change: Fixing Bush Internet in rural, regional, and remote Australia. In R. Hay, L. Eagle and A. Bhati (Eds.), *Broadening Cultural Horizons in Social Marketing*. Singapore: Springer. https://doi.org/10.1007/978-981-15-8517-3_13.

Hedley, C. (2015). The role of precision agriculture for improved nutrient management on farms. *Journal of the Science of Food and Agriculture* 95(1), 12–19. https://doi.org/10.1002/jsfa.6734.

Henry, B., Charmley, E., Eckard, R., Gaughan, J. B. and Hegarty, R. (2012). Livestock production in a changing climate: Adaptation and mitigation research in Australia. *Crop and Pasture Science* 63(3), 191–202.

Holmes, J., Burke, J., Campbell, L. and Hamilton, A. (2020). Towards a national broadband strategy for Australia 2020–2030. *Journal of Telecommunications and the Digital Economy* 8(4), 192–269. https://doi.org/10.18080/jtde.v8n4.371.

Hossain, M. E., Kabira, M. A., Zheng, L., Swain, D. L., McGrath, S. and Medway, J. (2022). A systematic review of machine learning techniques for cattle identification: Datasets, methods and future directions. *Artificial Intelligence in Agriculture* 6, 138–155. https://doi.org/10.1016/j.aiia.2022.09.002.

Hughes, N., Soh, W. Y., Lawson, K. and Lu, M. (2022). Improving the performance of micro-simulation models with machine learning: The case of Australian farms. *Economic Modelling* 115. https://doi.org/10.1016/j.econmod.2022.105957.

Jackson, E. L. and Cook, S. (2022). The complex adoption pathways of digital technology in Australian livestock supply chains systems. *Crop and Pasture Science* 74(6), 538-554. https://doi.org/10.1071/CP21593.

Kearton, T., Marini, D., Cowley, F., Belson, S. and Lee, C. (2019). The effect of virtual fencing stimuli on stress responses and behavior in sheep. *Animals* 9(1), 30.

Lamb, D. W. (Eds.) (2013). Proceedings of the inaugural digital rural futures conference, Smart Farms - Smart Regions. University of New England, Australia.

Lamb, D. W. (2017a). Accelerating precision agriculture to decision agriculture: A review of on-farm telecommunications challenges and opportunities in supporting a digital agriculture future for Australia. University of New England and Cotton Research and Development Corporation, September 2017, 128 pages. ISBN 978-1-921597-74-9 (Hard copy), ISBN 978-1-921597-75-6 (Digital).

Lamb, D. W. (2017b). Telecommunications and Australian agriculture: Will top down meet bottom up? *Australian Farm Policy Journal* 14(3), 31–47.

Lamb, D. W. (2018). Connecting Australian farmers: Gateways, gatekeepers and a forgotten solution. *Australian Farm Policy Journal* 15(1), 51–59.

Lamb, D. W., Frazier, P. and Adams, P. (2008). Improving pathways to adoption: Putting the right P's in precision agriculture. *Computers and Electronics in Agriculture* 61(1), 4-9.

Lamb, D. W., Taylor, D., Trotter, M., Donald, G. and Schneider, D. (2010). Precision pastures: Opportunities and challenges for spatial information to improve productivity and animal welfare in extensive livestock systems. In Proceedings of the 14th Symposium on Precision Agriculture in Australasia, Albury, NSW.

Lee, C. and Campbell, D. L. M. (2021). A multi-disciplinary approach to assess the welfare impacts of a new virtual fencing technology. *Frontiers in Veterinary Science* 8, 637709.

Lee, C., Henshall, J. M., Wark, T. J., Crossman, C. C., Reed, M. T., Brewer, H. G., O'Grady, J. and Fisher, A. D. (2009). Associative learning by cattle to enable effective and ethical virtual fences. *Applied Animal Behaviour Science* 119(1-2), 15-22.

Lee, C., Colditz, I. G. and Campbell, D. L. M. (2018). A framework to assess the impact of new animal management technologies on welfare: A case study of virtual fencing. *Frontiers in Veterinary Science* 5, 187. https://doi.org/10.3389/fvets.2018.00187.

Leigo, S., Brennan, G., Beutel, T., Gray, A., Phelps, D., Driver, T. and Trotter, M. (2012). Overview of technology products for the beef industry of remote Australia. CRC-REP Working Paper CW009. Ninti One Limited, Alice Springs. ISBN: 978-1-74158-219-2. Available at: https://nintione.com.au/resource/CW009_PrecisionPastoralManagementTools_ProductReview.pdf. (Accessed May 2023).

Lemphane, N. J., Kotze, B. and Kuriakose, R. B. (2022). A review on current IoT-based pasture management systems and applications of digital twins in farming. In R. C. Poonia, V. Singh, D. Singh Jat, M. J. Diván and M. S. Khan (Eds.), *Proceedings of the third international conference on sustainable computing. Advances in intelligent systems and computing*, Vol. 1404. Singapore: Springer, 173-180. https://doi.org/10.1007/978-981-16-4538-9_18.

Leonard, E. (Ed.), Rainbow, R. (Ed.), Trindall, J. (Ed.), Baker, I., Barry, S., Darragh, L., Darnell, R., George, A., Heath, R., Jakku, E., Laurie, A., Lamb, D., Llewellyn, R., Perrett, E., Sanderson, J., Skinner, A., Stollery, T., Wiseman, L., Wood, G. and Zhang, A. (2017). Accelerating precision agriculture to decision agriculture: Enabling digital agriculture in Australia. Cotton Research and Development Corporation, Australia. https://www.crdc.com.au/sites/default/files/CRD18001-001%20CRDC%20P2D%20Report%20low%20res.pdf. (Accessed March 2023).

Lockie, S., Fairley-Grenot, K., Ankeny, R., Botterill, L., Howlett, B., Mcbratney, A., Probyn, E., Sorrell, T., Sukkarieh, S. and Woodhead, I. (2020). The future of agricultural technologies. Australian Council of Learned Academies (ACOLA).

Long, J., Takahata, H., Umetsu, K., Hoshiba, H. and Takeyama, I. (1991). A livestock walk-through scale system. *Nogyo Shisetsu (Journal of the Society of Agricultural Structures, Japan)* 21(3), 175-182.

Lowder, S. K., Skoet, J. and Raney, T. (2016). The number, size, and distribution of farms, smallholder farms, and family farms worldwide. *World Development* 87, 16-29.

Lyons, N. A. and Lomax, S. (2022). Developments in activity and location technologies for monitoring cattle movement and behaviour. In D. Berckmans (Ed.), *Advances in Precision Livestock Farming*. Belgium: Burleigh Dodds Science Publishing Limited, 123-148. https://doi.org/10.19103/AS.2021.0090.05.

Manning, J. K., Fogarty, E. S., Trotter, M. G., Schneider, D. A., Thomson, P. C., Bush, R. D. and Cronin, G. M. (2014). A pilot study into the use of global navigation satellite system technology to quantify the behavioural responses of sheep during simulated dog predation events. *Animal Production Science* 54(10), 1676–1681. https://doi.org/10.1071/AN14221.

Manning, J., Cronin, G., González, L., Hall, E., Merchant, A. and Ingram, L. (2017). The behavioural responses of beef cattle (Bos taurus) to declining pasture availability and the use of GNSS technology to determine grazing preference. *Agriculture* 7(5), 45.

Manning, J., Power, D. and Cosby, A. (2021). Legal complexities of animal welfare in Australia: Do on-animal sensors offer a future option? *Animals* 11(1), 91.

Mark, T. B., Griffin, T. W. and Whitacre, B. E. (2016). The role of wireless broadband connectivity on 'Big Data' and the agricultural industry in the United States and Australia. *International Food and Agribusiness Management Review* 19(1030-2016-83150), 43–56.

Marshall, A., Dezuanni, M., Burgess, J., Thomas, J. and Wilson, C. K. (2020). Australian farmers left behind in the digital economy–Insights from the Australian Digital Inclusion Index. *Journal of Rural Studies* 80, 195–210.

Mayberry, D., Bartlett, H., Moss, J., Davison, T. and Herrero, M. (2019). Pathways to carbon-neutrality for the Australian red meat sector. *Agricultural Systems* 175, 13–21.

McFadden, J., Casalini, F., Griffin, T. and Antón, J. (2022). The digitalisation of agriculture: A literature review and emerging policy issues. OECD Food, Agriculture and Fisheries Papers, No. 176. Paris: OECD Publishing. https://doi.org/10.1787/285cc27d-en.

McKinna, D. and Wall, C. (2020). Commercial application of supply chain integrity and shelf life systems. Final Report V.MFS.0447 (Meat and Livestock Australia Limited: North Sydney). Available at: https://www.mla.com.au/globalassets/mla-corporate/research-and-development/final-reports/2020/v.mfs.0447-final-report.pdf. (Accessed May 2023).

Meat and Livestock Australia (2021). Assessing the feasibility of an implantable ID for cattle. Summary Report V.RDA.0002 (Meat and Livestock Australia Limited: North Sydney). Available at: https://www.mla.com.au/globalassets/mla-corporate/research-and-development/final-reports/2021/project-summary-v.pdf. (Accessed May 2023).

Meat and Livestock Australia (2023a). MLA Australian feedbase monitor. Available at: https://www.mla.com.au/extension-training-and-tools/tools-calculators/australian-feedbase-monitor/. (Accessed May 2023).

Meat and Livestock Australia (2023b). Soil, stock and carbon trifecta. Available at: https://www.mla.com.au/news-and-events/industry-news/soil-stock-and-carbon-trifecta/. (Accessed May 2023).

Meehan, D. (2019). Evaluation of connectivity and digital solutions for vertically integrated beef operations – Stages 1 and 2. Final Report P.PSH.1056 (Meat and Livestock Australia Limited: North Sydney). Available at: https://www.mla.com.au/contentassets/e4ca7a01dff9412d95dc4cd61cd53b5c/p.psh.1056_final_report.pdf. (Accessed May 2023).

Menzies, D., Patison, K. P., Corbet, N. J. and Swain, D. L. (2018). Using Walk-over-weighing technology for parturition date determination in beef cattle. *Animal Production Science* 58(9), 1743–1750. https://doi.org/10.1071/AN16694.

Milsom, A., Bell, O., Bailey, K., Brown, S. C., Barton, R. A., Moreno Garcia, C. A., Chen, A., Bryant, R. H., Maxwell, T. M. R. and Eady, C. C. (2019). Assessing the ability of a stationary pasture height sensing device to estimate pasture growth and biomass.

Journal of New Zealand Grasslands 81, 61–68. https://doi.org/10.33584/jnzg.2019.81.384.

Moore, A. D., Bell, L. W. and Revell, D. K. (2009). Feed gaps in mixed-farming systems: Insights from the Grain & Graze program. *Animal Production Science* 49(10), 736–748. https://doi.org/10.1071/AN09010.

Moore, G. A., Sanford, P., Dolling, P. J. and Real, D. (2021). The challenges of developing resilient perennial pastures for a Mediterranean environment–a review for Western Australia. *Crop and Pasture Science* 72(9), 613–633. https://doi.org/10.1071/CP20304.

Mulrooney, K., Harkness, A., Barwick, J., Nolan, H. and Schneider, D. (2022). Ceres Teg: An evaluation for the prevention, interruption and reduction of livestock theft. Final Report. University of New England Centre for Rural Crimminology (University of New England Australia). 26pp. Available at: https://cpb-ap-se2.wpmucdn.com/blog.une.edu.au/dist/3/1351/files/2022/09/Ceres-Tag-An-Evaluation-for-the-Prevention-Interruption-and-Reduction-of-Livestock-Theft.pdf. (Accessed May 2023).

Mundava, C., Schut, A. G. T., Helmholz, P., Stovold, R., Donald, G. and Lamb, D. W. (2015). A novel protocol for assessment of above ground biomass in rangeland environments. *Rangeland Journal* 37(2), 157–167. https://doi.org/ 10.1071/RJ14072.

NBNCo (2006). Kirby 'smart' farm. Available at: https://www.nbnco.com.au/corporate-information/media-centre/image-gallery/kirby-smart-farm. (Accessed April 2023).

Neethirajan, S. and Kemp, B. (2021). Digital livestock farming. *Sensing and Bio-Sensing Research* 32, 100408. https://doi.org/10.1016/j.sbsr.2021.100408.

Perrett, E., Heath, R., Laurie, A. and Darragh, L. (2017). Accelerating precision agriculture to decision agriculture – analysis of the economic benefit and strategies for delivery of digital agriculture in Australia. Australian Farm Institute and Cotton Research and Development Corporation. Available at: https://www.crdc.com.au/sites/default/files/P2D%20Ecomomic%20impact%20of%20digital%20ag%20-%20AFI%20Final%20Report.pdf. (Accessed March 2023).

Pringle, M. J., O'Reagain, P. J., Stone, G. S., Carter, J. O., Orton, T. G. and Bushell, J. J. (2021). Using remote sensing to forecast forage quality for cattle in the dry savannas of northeast Australia. *Ecological Indicators* 133, 108426.

Putfarken, D., Dengler, J., Lehmann, S. and Härdtle, W. (2008). Site use of grazing cattle and sheep in a large-scale pasture landscape: A GPS/GIS assessment. *Applied Animal Behaviour Science* 111(1–2), 54–67.

Rahman, M. M. and Lamb, D. W. (2017). The role of directional LAI in determining the fAPAR–NDVI relationship when using active optical sensors in tall fescue (Festuca arundinacea) pasture. *International Journal of Remote Sensing* 38(11), 3219–3235. https://doi.org/10.1080/01431161.2017.1292069.

Rahman, M. M., Lamb, D. W., Stanley, J. N. and Trotter, M. G. (2014). Use of proximal sensors to evaluate at the sub-paddock scale a pasture growth-rate model based on light-use efficiency. *Crop and Pasture Science* 65(4), 400–409. https://doi.org/10.1071/CP14071.

RDA (2014). UNE Smart Farm named Northern Inland Innovation of the Year. Available at: https://www.rdani.org.au/news-events/newsletters/une-smart-farm-named-northern-inland-innovation-of-the-year-10-11-2014.php. (Accessed April 2023).

Roberts, J., Trotter, M., Schneider, D., Lamb, D., Hinch, G. and Dobos, R. (2015). Daily grazing time of free-ranging cattle as an indicator of available feed. In Proceedings of the 7th European Conference on Precision Livestock Farming.
Rotz, S., Duncan, E., Small, M., Botschner, J., Dara, R., Mosby, I., Reed, M. and Fraser, E. D. G. (2019). The politics of digital agricultural technologies: A preliminary review. *Sociologia Ruralis* 59(2), 203–229. https://doi.org/ 10.1111/soru.12233.
Schaefer, M. T., Lamb, D. W. and Bradbury, R. (2014). A comparison of two ranging approaches in an active, optical plant canopy sensor. IEEE Sensors Applications Symposium Proceedings, Queenstown, NZ, 253–257 (ISBN:978 1-4799-2180-5).
Schulz, P., Prior, J., Kahn, L. and Hinch, G. (2022). Exploring the role of smartphone apps for livestock farmers: Data management, extension and informed decision making. *The Journal of Agricultural Education and Extension* 28(1), 93–114.
Seitzinger, A., Garner, G., Bradhurst, R., Okelo, W. and Rodan, M. (2022). Analysis of livestock movements and the benefits and costs of livestock standstills in Australia as part of the response to foot-and-mouth disease (CSIRO, Australia). 114pp. Available at: https://www.agric.wa.gov.au/sites/gateway/files/DPIRD_CSIRO_NLSS%20cost%20benefit%20analysis.pdf. (Accessed May 2023).
Shepherd, M., Turner, J. A., Small, B. and Wheeler, D. (2020). Priorities for science to overcome hurdles thwarting the full promise of the 'digital agriculture' revolution. *Journal of the Science of Food and Agriculture* 100(14), 5083–5092. https://doi.org/10.1002/jsfa.9346.
Smith, R. A. and Turner, M. J. B. (1974). Electronic aids for use in fatstock weighing. *Journal of Agricultural Engineering Research* 19(3), 299–311.
Swain, D. L. and Bishop-Hurley, G. J. (2007). Using contact logging devices to explore animal affiliations: Quantifying cow–calf interactions. *Applied Animal Behaviour Science* 102(1–2), 1–11.
Tang, W., Biglari, A., Ebarb, R., Pickett, T., Smallidge, S. and Ward, M. (2021). A smart sensing system of water quality and intake monitoring for livestock and wild animals. *Sensors* 21(8), 2885. https://doi.org/10.3390/s21082885.
Taylor, D. B., Schneider, D. A., Brown, W. Y., Price, I. R., Trotter, M. G., Lamb, D. W. and Hinch, G. N. (2011). GPS observation of shelter utilisation by Merino ewes. *Animal Production Science* 51(8), 724–737. https://doi.org/10.1071/AN11025.
Taylor, K., Griffith, C., Lefort, L., Gaire, R., Compton, M., Wark, T., Lamb, D. W., Falzon, G. and Trotter, M. (2013). Farming the web of things. *IEEE Intelligent Systems* 28(6), 12–19. https://doi.org/10.1109/MIS.2013.102.
Thomas, D. T., Wilmot, M. G., Alchin, M. and Masters, D. G. (2008). Preliminary indications that Merino sheep graze different areas on cooler days in the Southern Rangelands of Western Australia. *Australian Journal of Experimental Agriculture* 48(7), 889–892.
Tobin, C., Bailey, D. W., Trotter, M. G. and O'Connor, L. (2020). Sensor based disease detection: A case study using accelerometers to recognize symptoms of Bovine Ephemeral Fever. *Computers and Electronics in Agriculture* 175, 1–5. https://doi.org/10.1016/j.compag.2020.105605.
Tobin, C., Bailey, D. W. and Trotter, M. G. (2021). Tracking and sensor-based detection of livestock water system failure: A case study simulation. *Rangeland Ecology and Management* 77, 9–16. https://doi.org/10.1016/j.rama.2021.02.013.
Tomkins, N. and O'Reagain, P. (2007). Global positioning systems indicate landscape preferences of cattle in the subtropical savannas. *The Rangeland Journal* 29(2), 217–222.

Trotter, M. G., Lamb, D. W. and Hinch, G. N. (2010a). GNSS tracking of livestock: Towards variable rate fertilizer strategies for the grazing industry. 10th International Conference on Precision Agriculture, 18–21 July, Denver.

Trotter, M. G., Lamb, D. W., Hinch, G. and Guppy, C. (2010b). Global navigation satellite systems (GNSS) livestock tracking: System development and data interpretation. *Animal Production Science* 50, 616–623. https://doi.org/10.1007/s12665-010-0582-7.

Trotter, M. G., Trotter, T. F. and Lamb, D. W. (Eds.) (2010c). Proceedings of the 1st Australian and New Zealand spatially enabled livestock management symposium. Precision Agriculture Research Group, University of New England, Australia. ISBN 978-1-921597-23-7.

Trotter, M. G., Lamb, D. W., Donald, G. E. and Schneider, D. A. (2010d). Evaluating an active optical sensor for quantifying and mapping green herbage mass and growth in a perennial grass pasture. *Crop and Pasture Science* 61(5), 389–398. https://doi.org/10.1071/CP10019.

Trotter, M. (2013). PA Innovations in livestock, grazing systems and rangeland management to improve landscape productivity and sustainability. *Agricultural Science* 25(1), 27–31.

Trotter, M. (2019). Precision livestock farming and pasture management systems. In J. Stafford (Ed.), *Precision Agriculture for Sustainability*. Cambridge: Burleigh Dodds, 421–459. Available at: https://www.taylorfrancis.com/. (Accessed May 2023).

Trotter, M., Cosby, A., Manning, J., Thomson, M., Trotter, T., Graz, P., Fogarty, E., Lobb, A. and Smart, A. (2018). Demonstrating the value of animal location and behaviour data in the red meat value chain. Final Report P. PSH.0835 (Meat and Livestock Australia Limited: North Sydney). Available at: https://www.mla.com.au/research-and-development/reports/2018/demonstrating-the-value-of-animal-location-and-behaviour-data-in-the-red-meat-value-chain/. (Accessed May 2023).

Trotter, M., Bailey, D., Manning, J., Evans, C., Fleury Azevedo Costa, D., Fogarty, E. and Chang, A. (2022). Developments in on-animal sensors for monitoring livestock. In D. Berckmans (Ed.), *Advances in Precision Livestock Farming*. Belgium: Burleigh Dodds Science Publishing Limited. https://doi.org/10.19103/AS.2021.0090.01.

von Keyserlingk, M. A. G., Phillips, C. J. C. and Nielsen, B. L. (2016). Water and the welfare of farm animals. In C. Phillips (Ed.), *Nutrition and the Welfare of Farm Animals. Animal Welfare*, Vol. 16. Cham: Springer, 183–197. https://doi.org/10.1007/978-3-319-27356-3_9.

Wagner, J. J. and Engle, T. E. (2021). Invited review: Water consumption, and drinking behavior of beef cattle, and effects of water quality. *Applied Animal Science* 37(4), 418–435. https://doi.org/10.15232/aas.2021-02136.

Waters, P. and Koch, A. (2022). The broadband futures forum: The Australian broadband advisory council Agri-Tech expert working group report. *Journal of Telecommunications and the Digital Economy* 10(1), 23–33.

Wheeler, J., Hedges, D. and Mulcahy, C. (1977). The use of dyebanding for measuring wool production and fleece tip wear in rugged and unrugged sheep. *Crop and Pasture Science* 28(4), 721–735.

Williams, L. R., Moore, S. T., Bishop-Hurley, G. J. and Swain, D. L. (2020). A sensor-based solution to monitor grazing cattle drinking behaviour and water intake. *Computers and Electronics in Agriculture* 168, 105141.

Williams, T., Wilson, C., Wynn, P. and Costa, D. (2021). Opportunities for precision livestock management in the face of climate change: A focus on extensive systems. *Animal Frontiers* 11(5), 63–68.

Wilson, C. S., Jenkins, D. J., Brookes, V. J. and Barnes, T. S. (2019). An eight-year retrospective study of hydatid disease (Echinococcus granulosus sensu stricto) in beef cattle slaughtered at an Australian abattoir. *Preventive Veterinary Medicine* 173, 104806.

Windsor, P. (2021). Progress with livestock welfare in extensive production systems: Lessons from Australia. *Frontiers in Veterinary Science* 8-2021. https://doi.org/10.3389/fvets.2021.674482.

Zhou, I., Lipman, J., Abolhasan, M., Shariati, N. and Lamb, D. W. (2020). Frost monitoring cyber-physical system: A survey on prediction and active protection methods. *IEEE Internet of Things Journal* 7(7), 6514–6527. https://doi.org/10.1109/JIOT.2020.2972936.

Index

www.ingramcontent.com/pod-product-compliance
Ingram Content Group UK Ltd.
Pitfield, Milton Keynes, MK11 3LW, UK
UKHW021826100926
8556IPUK00003B/188

9 781801 463829